Laws and Models
in Science

Laws and Models
in Science
edited by Donald Gillies

ISBN 0-9543006-61
King's College Publications
Scientific Director: Dov Gabbay
Managing Director: Jane Spurr
Department of Computer Science
Strand, London WC2R 2LS, UK
kcp@dcs.kcl.ac.uk
www.dcs.kcl.ac.uk/kcl-publications/

Cover design by Richard Fraser, www.avalonarts.co.uk
Printed by Lightning Source, Milton Keynes, UK

CONTENTS

Introduction   v
Donald Gillies

Knowledge Gain and Practical Use: Models in Pure and   1
Applied Science
Martin Carrier

Opinion Dynamics: Insights by Radically Simplifying   19
Models
Rainer Hegselmann

A Forgotten Common Origin: Comments on Hegselmann   47
Moshé Machover

Models and Modals   51
Huw Price

Republican Perspectivalism: A Comment on Huw Price   73
Raffaella Campaner

Truth and the Aim of Belief   79
Pascal Engel

Mechanisms of Truth-directedness: Comments on Pascal   101
Engel's 'Truth and Aim of Belief'
Wlodek Rabinowicz

Beliefs as Mental States and as Actions: A Comment on   107
Pascal Engel
Peter Rosner

How a Type-type Identity Theorist Can Be a   115
Non-reductionist: An Answer from the Idealizational
Conception of Science
Katarzyna Paprzycka

On the Epistemic Significance of Type–Type Identities: A   131
Comment on Katarzyna Paprzycka
Marcel Weber

Response to Marcel Weber's Comment   135
Katarzyna Paprzycka

Idealisation and Mathematisation in Cassirer's Critical   139
Idealism
Thomas Mormann

Cassirer's Critical Idealism: A Comment on Thomas   161
Mormann
Maarten van Dyck and Erik Weber

Recursive Causality in Bayesian Networks and Self-fibring   173
Networks
Jon Williamson and Dov Gabbay

Recursive Causality and the Causal Relata: Comments on   223
Williamson and Gabbay
David Glass

On Gabbay's Fibring Methodology for Bayesian and Neural 233
Networks: A Comment on Williamson and Gabbay
Artur D'Avila Garcez

Philosophy of Science in a European Perspective   247
Maria Carla Galavotti

Index   253

Introduction

DONALD GILLIES

1 Introduction to the Introduction

The present volume has its origin in a conference held at King's College London, 7–9 September 2003, on Laws and Models in Science. This conference was the third and last in a series organised by the European Science Foundation Network on Historical and Contemporary Perspectives of Philosophy of Science in Europe. The volume does not exactly correspond to the conference. Some of the contributors who gave talks were unable to provide a written version, and conversely one of the commentators who could not attend the conference was able to contribute a piece to the volume. Moreover those speakers who did prepare a written version had time to revise their papers. Despite these divergences, the papers are here given in the order in which they were presented at the conference.

As well as the papers and comments, the volume contains a useful overview of the activities of the European Science Foundation Network on Historical and Contemporary Perspectives of Philosophy of Science in Europe. This raises the interesting question of whether a distinctive approach to philosophy of science is likely to emerge in Europe. I will return to this issue at the end of this introduction. At one point in her overview, Maria Carla Galavotti compares the present situation of philosophy of science with its situation in the days of the Vienna Circle. Her point is that the present network agrees with the Vienna Circle (p. 249)[1] 'in regarding the dialogue between philosophers and scientists as indispensable'. However the situation in science is now very different. The period of the Vienna Circle overlapped with the early twentieth century revolution in physics, and the main foci of the Circle's discussions of science were relativity and quantum mechanics. These topics are still important for philosophy of science today, but there are many new fields emerging in today's science which have not yet been analysed to any great extent by philosophers of science. As Galavotti says (p.248): 'a main concern ... has been to bring together philosophers and scientists, especially from areas like economics, psychology, brain research, artificial intelligence, which are more "new" to philosophy of science than other disciplines, like physics, traditionally involved with foundational problems.' Two of these new areas for philosophy of science, namely psychology and artificial intelligence are well represented in

Laws and Models in Science, v–xiv.
© 2004, the author.

the present volume.

The new discoveries in physics which were analysed by the Vienna Circle mainly took the form of general laws expressed by mathematical equations such as the Lorentz transformation, the field equations of General Relativity, and the Schrödinger equation. The new developments in the science of today have brought different approaches. As the title of this volume suggests there is an increasing emphasis on models as well as laws, and this has gone with a greater use of causality and statistical methods.

2   An Overview of the Use of Models

Martin Carrier's paper is a very appropriate one with which to open the collection because it gives subtle analysis of the role of local models in pure and applied research. Carrier considers application-dominated research and gives interesting examples of very local models. One of these concerns the use of 'giant magnetoresistance' to produce magnetic read heads. The design rules involved are so specific that those (p.8) 'used by Philips don't work for IBM read heads.' Examples like this suggest that (p. 10) 'applied research is methodologically inferior to epistemic science.' However Carrier argues against this conclusion by showing that local models can be fruitfully used in epistemic research as well. A nice example is Prandtl's account of the flow of viscous fluid. This flow is governed by the Navier-Stokes equation which unfortunately is unsolvable. Prandtl overcame the problem by dividing the fluid into a thin layer of very high viscosity near the surface of bodies and the rest which was regarded as of negligible viscosity. These assumptions, though strictly unrealistic, produce an empirically adequate solution of the equation. Moreover local models, such as Prandtl's, can sometimes (p. 13) 'provide a mechanism and some sort of causal understanding of the processes involved in fluid flow', whereas calculation from equations may yield (p. 13) 'numbers but no understanding.' Carrier's general conclusion is that (p. 15): 'Epistemic and applied research share important methodological characteristics.'

3   Some General Philosophical Themes

The present volume contains both papers concerned with general philosophical themes and also papers dealing with quite specific models. The two sorts of paper are connected, however, because the specific models are dealing with some of the issues raised by the general themes. The general themes include: probability and causality (Price), knowledge and belief (Engel), reductionism and the mind-body problem (Paprzycka). The specific models concern the formation of belief (Hegselmann), and Bayesian and neural networks in artificial intelligence (Gabbay and Williamson). The links between the general themes and these specific models are clear enough. In this section I will discuss the treatment of the general themes.

Huw Price begins his paper with Russell's famous argument that causality, like the monarchy, should be eliminated. He sees this as posing the false dichotomy that either causes exist in the world or they should be eliminated. Price claims that, extending Russell's metaphor, this is like forcing someone to choose between monarchy or anarchy. There is, in reality, an intermediate possibility, namely republicanism. Thus one could speak of a causal republicanism. This is a special case of Price's advocacy of (p. 51) 'a kind of pragmatic perspectivalism about modal notions.' In this paper he applies this to probabilistic models. These are usually interpreted as either objective or subjective, but Price argues for a third, intermediate, possibility: perspectival models. On this account (p. 62), 'Probability passes muster as a notion we ourselves bring to our models of the world — a possibility that requires that we see ourselves as Kant does, as conceptual creators, not as mere consumers of conceptual categories pre-packaged by God.' Huw Price's commentator, Raffaella Campaner, argues that perpectivalism may be a matter of degree, and that more work is needed to extend it to other modal notions.

Pascal Engel deals in his paper with the central notions of truth and belief. He considers the thesis that truth is the aim of belief, and gives three interpretations of this supposed truth directedness of belief – the causal, the intentional, and the normative. The causal account he regards as correct but incomplete. It is not sufficient to characterise belief properly. The intentional account, however, he rejects principally on the grounds that (p. 89) 'it wrongly equates believing to purposive attitudes such as guessing.' Engel is most sympathetic to the normative account, but he modifies it by suggesting that the norm which governs belief should be knowledge rather than truth. Most of the paper is concerned with full beliefs, but, in an appendix, Engel considers what modifications might be needed if one adopts a Bayesian approach and limits oneself to partial beliefs.

There are two commentators on Engel's paper. Wlodek Rabinowicz questions (p. 99) 'Pascal Engel's scepticism concerning the possibility of interpreting the truth-directedness of belief in causal-functional terms.' Rabinowicz regards Velleman as holding a causal-functional account of truth-directedness as opposed to the intentional account which Engel attributes to him. It might be objected to such a causal-functional account (as Engel does to the intentional account) that it applies equally to guessing and so is not distinctive enough. However, Rabinowicz does not think that (p. 103) 'this objection is especially worrying. To begin with, why should we require that belief must be the only kind of attitude that is truth-directed? Surely, guessing is another such attitude ... in a well-functioning agent, guessing is regulated by mechanisms that are less exacting than those that regulate belief: For guessing, the expected degree of truth-tracking is lower. Which means that, in this causal-functional sense, one can distinguish the more exacting kind of truth-directedness that characterizes belief from the less exacting one that characterizes guessing.'

The second commentator, Peter Rosner, raises two different points concerning

Pascal Engel's paper. The first is that intentional actions do play a part in belief formation. For example, a later belief is often a revision of an earlier belief and this revision may be the result of the intention of being prepared to look out for evidence which contradicts existing beliefs. Moreover (p. 107): '...even if one accepts that beliefs are not under voluntary control .. it is under control, at least to some extent, whether one forms a belief or not.' Rosner's example here is belief about the Dollar/Euro parity. Some people may try to form a belief about this, while others may not bother to do so. Rosner's second point is that Engel considers only beliefs in simple propositions. He invites us to consider also beliefs like the following (p. 109): '*I believe that, all things considered, detailed regulations of the economy have at best no results. It is better to leave markets alone.*' Such fundamental beliefs which express a *Weltanschauung* are not like simple empirical beliefs, but they are not mere feelings either, since reasons for and against such beliefs are given. Characteristically, however, economists continue to disagree about such fundamental propositions.

Katarzyna Paprzycka's paper starts with the classic problem of whether the mind can be reduced to the body. Perhaps the most well-known recent approach is Davidson's anomalous monism which tries to establish a non-reductive material- ism by claiming that there are token-token identities but not type-type identities be- tween mental and physical events. First Stoutland in 1976 and later others have ob- jected to Davidson's theory on the grounds that it leads to type-epiphenomenalism since, without type-type identities, (p. 113) 'it becomes unclear how the mental as mental can be viewed as being causally efficacious at all.' Paprzycka applies some ideas from the philosophy of science, namely the idealizational conception of science, to this problem. The key point about the idealizational conception of science is that it considers a theory as having the structure of an idealizational law which can be corrected by a succession of secondary factors. A classic example is the ideal gas law which can be modified by secondary factors to produce van der Waals' equation. If we apply this to the reduction of one theory t to another T, two cases become possible. First of all the coordinating definitions establish an isomorphism between the idealizational laws and the succession of secondary factors. This occurs (p.122) in the case of 'the reduction of qualitative thermody- namics to statistical thermodynamics, where both the idealizational law (the ideal gas law) and its concretizations (van der Waals' corrections) can be derived from statistical thermodynamics.' Such theories are said to be essentially compatible. Secondly it is possible also to have a situation where the coordinating definitions do not preserve the isomorphism. This gives a pair of essentially incompatible theories. Such theories can give us type-type identities without reduction. So in effect producing a new version of Davidson's approach in which the threat of type- epiphenomenalism is avoided. It actually appears not unlikely that there might be type-type identities between thoughts and brain-processes while the corresponding psychological and neuro-physiological theories are indeed incompatible.

Marcel Weber makes some interesting criticisms of Paprzycka's paper to which she replies. It will be convenient to consider Weber's comments and Paprzycka's response together. It will also be convenient to take Weber's two points in reverse order. Weber's second point is to criticize Paprzycka's use of Nagel's derivational account of reduction. He says (p. 132): 'In the philosophy of biology, for example, it was shown that the laws of classical genetics are not derivable from the principles of molecular biology ... However, non-derivational accounts of reduction can accommodate the case ...' Paprzycka in her response accepts this point, saying (p. 133): '... I by no means think that derivability is essential to reduction'. Moreover she hopes to deal with this question in detail in a larger project of which her present paper is a part.

Weber also criticizes Nagel's approach on the grounds that one needs type-type identities which are stronger than Nagel's bridge laws. Weber thinks that Paprzycka operates with equalities rather than type-type identities proper, and that if one strengthens her equalities to identities her case of type-type identities between incompatible theories becomes implausible. Paprzycka replies (p. 133): ' I *do* mean factor identities to be factor identities and not just equalities ...', and she goes on to give a rather more complicated example to illustrate the possibility of type-type identities between essentially incompatible theories. This is a vigorous and stimulating debate.

4 A Role for History

According to Galavotti (p. 251): 'A further characteristic of the European approach to philosophy of science lies with its historical concern.' In this collection, historical interests are well demonstrated in Thomas Mormann's paper on Cassirer. In this paper, Mormann laments the neglect of Cassirer's Critical Idealism among contemporary philosophers of science. One thesis of Cassirer's which Mormann finds particularly interesting is the 'sameness thesis' that the concepts of mathematics and the concepts of the empirical sciences are of the same kind, or, more precisely, that (p. 142) *'the common foundational syntheses on which mathematical and physical knowledge are based, are provided by the method of "ideal elements".'* This stress on idealization relates Cassirer to the idealizational conception of science which Paprzycka used in her paper. However, Mormann points out (p. 141) that the idealizational conception of Leszek Nowak and his school only considers the empirical sciences while Cassirer extends the conception to mathematics. Indeed 'ideal elements' are often used in mathematics to complete a domain. For example, complex numbers are a completion of the domain of reals, while Dedekind cuts can be used to create a domain of reals which completes the domain of rationals. Now in physics too there are ideal elements such as points, lines, the elliptical orbits of planets, etc. Whitehead had a programme to construct ponts out of regions of space. Mormann comments (p. 150): 'The interesting point is that this construction of points from regions can be conceived as a generalisation of Dedekind's method of cuts.' This shows the close analogy between ideal elements in mathematics and physics. Mormann suggests that Cassirer's critical idealism might be considered as (p. 155) 'a kind of moderate conventionalism.'

In their comment on Mormann's paper, Maarten Van Dyck and Erik Weber sympathize with Mormann's defence of Cassirer in general terms, but they cast doubt on classifying Cassirer's position as 'moderate conventionalism' by comparing Cassirer to Poincaré. Van Dyck and Weber argues that there are notable differences between these two thinkers. For example, Poincaré held that geometry and arithmetic were different. Conventionalism applied to geometry, but arithmetic was synthetic *a priori* because of the principle of mathematical induction. Cassirer, on the other hand, gives the same status to arithmetic as to geometry. Moreover, in general, Van Dyck and Weber say (p. 167): 'We believe that Cassirer's attention to the dynamic character of thought is one of the aspects of his thought that clearly distinguishes him from Poincaré's conventionalism.'

This last point of Van Dyck and Weber is a good one, and may also help to explain the neglect of Cassirer's work in the philosophy of mathematics. It must be remembered that this work of Cassirer's is not just neglected by contemporary philosophers, but was neglected by philosophers of mathematics at the time. Now when Cassirer was writing in the first few decades of the 20th century, philosophy of mathematics was dominated by the non-historical foundational schools. It was not till after the work of Lakatos in the 1960s that a historical approach was introduced into the philosophy of mathematics, and even today it is a minority current compared with attempts to revive logicism, formalism and constructivism. In Cassirer's own day, a historical approach to the philosophy of mathematics must have seemed rather bizarre, and it is perhaps this feature of Cassirer's work, rather than his critical idealism, which led to its neglect. However, this same feature may lead to a revival of interest in Cassirer, now that a historical approach to the philosophy of mathematics has gained some adherents.

5   Some specific models

Rainer Hegselmann's paper is concerned with the problem of how members of a group, initially holding different opinions, come to change their opinions in the light of the opinions of other members of the group. To explore this 'opinion dynamics', Hegselmann introduces the concept of 'a radically simplifying model', that is (p. 19) 'a model based on a very small set of assumptions which are "easy to handle".' One such model is the 'bounded confidence model'. This lends itself to computer simulation, and Hegselmann shows with a sequence of appealing graphical illustrations how this can lead either to a plurality of opinions, or to polarisation into two main opinions, or to consensus. His investigation continues by replacing the original simplifying assumptions by more complicated ones and examining the effect. This is a beautiful example of how the radically simplifying approach combined with computer simulations can throw light on a complex problem. Hegselmann seems to regard his bounded confidence model as too simple to have realistic applications and yet his assumptions come remarkably close to those made by Keynes in explaining investment decisions made by a community of entrepreneurs under conditions of uncertainty (see Keynes, 1936, Chapter 12, pp. 147–164, and Keynes, 1937, Section II, pp.112–119).

Moshé Machover in his comments on Hegselmann's paper draws attention to a largely forgotten historical contribution to the study of opinion dynamics. This is Lionel Penrose's 1952 booklet of only 73 pages entitled *On the Objective Study of Crowd Behaviour*. This work, though generally ignored both at the time of its publication and for a long period afterwards, contains many ideas which were later reinvented. These include a suggested measure of voting power, and contributions to the epidemiology of ideas which may yet prove valuable. One unexpected result of the conference will perhaps be to encourage the study of two important but neglected authors: Ernst Cassirer and Lionel Penrose.

The paper by Jon Williamson and Dov Gabbay discusses some models which are designed for use in artificial intelligence (AI). One of the most common ways of modelling causality in AI is the use of Bayesian networks. In the first part of the present paper, Bayesian networks are generalised to handle recursive causality which arises when a causal relation is part of another causal relation. For example smoking causes cancer (SC) is itself a causal relation, but this causal relation can cause governments to restrict tobacco advertising (A) so that SC causes A. If we allow such recursive causality, Bayesian networks need to be generalised to recursive Bayesian networks, and the first five sections of the paper show how this can be done. Section 7 shows how an analogous generalisation can be carried out for another type of causal model: a structural equation model. Up to this point, the paper has considered models for causality, but networks can also be used to model non-causal situations such as argumentation and information. The second part of the paper shows how the recursive generalisation can be applied to argumentation and information networks to produce what are called self-fibring networks, which have interesting applications in logic.

In his comments, David Glass subjects the notion of recursive causality to a careful analysis. He considers whether the recursive element can be eliminated from recursive Bayesian networks in three cases: human actions, prevention, and pre-emption. In the first two cases elimination seems to be possible, but the third case (pre-emption) does definitely seem to be better handled by recursive causality than by alternative approaches involving intermediate variables. David Glass points out that this defence of recursive causality has a wider significance, since acceptance of recursive causality seems to involve rejection of the event view of causality.

In his comment Artur Garcez shows how the fibring approach can be extended from Bayesian networks to neural networks. He illustrates his approach by applying it to an example of argumentation in legal reasoning and concludes with a strong advocacy of Gabbay's fibring methodology.

The various specific models in the papers discussed in this section have obviously only been made possible by the development of the computer. Computer modelling is indeed a new scientific technique, and its introduction has also changed the subject matter of philosophy of science from what it was in the days of the Vienna Circle.

6   A European Approach to Philosophy of Science?

Having reviewed the papers of the present volume, let us return to the intriguing question of whether a distinctive European approach to philosophy of science may emerge. There have, after all, recently been political differences between Europe and the USA over the Iraq war. So if there can be differences in politics, why not in philosophy of science as well? Against such a view, however, it could be argued that we are living now in the age of globalisation, and so the location of particular philosophical schools in particularly regions may be a thing of the past. Consider, for example, those two great schools of general philosophy — analytic philosophy and 'continental' philosophy. Analytic philosophy is supposed to flourish in the English-speaking world and continental philosophy in the European continent. In fact, however, there are many analytic philosophers living in the continent of Europe and many continental philosophers in Australia, Canada, the UK and the USA. Moreover technological advances of recent times such as email, the internet, ease of travel to international conferences, etc. all, so it could be argued, make it unlikely that schools of philosophy will be located in particular geographical areas. In addition to all this, academics from one country frequently get jobs in another country. A procedure which is again likely to extend schools of philosophy globally.

An example will illustrate the forces which are at work. In 1992 a collection which I had edited on Revolutions in Mathematics was published by Oxford University Press. The use of email and the internet was not so widespread in those days as it is now, and yet the variety of nationalities of the 12 contributors is already quite striking. To judge by the place of abode at the time of publication, there were 4 Americans, 1 Chinese, 3 English, 3 Germans and 1 Italian. In fact, however, one of the 'Americans' was an Italian who had just got a job in the USA, while one of the 'Germans' was an Italian who now teaches in Paris. It surely seems plausible that those interested in history and philosophy of mathematics will in future form a global network communicating by email and meeting at international conferences from time to time. Thus this particular school or sub-school of philosophy is unlikely to be located in a particular geographical region.

Such globalising tendencies are undoubtedly real, but perhaps they are not the whole story. Most schools of philosophy will, I am sure, spread out globally, but this does not preclude the possibility of some schools being more strongly represented in some areas than in others. Since the political climate and cultural traditions have an influence on academic appointments, we would indeed expect to find some schools or approaches more strongly represented in some countries than in others. In this sense then I think we may expect to find some tendencies more marked in European philosophy than, for example, in American philosophy. The analogy to political differences over the Iraq war is helpful here. It was not the case that everyone in Europe was opposed to the war. In fact many Europeans strongly supported the war. Conversely many Americans were strongly against the war. Yet there was still a difference. The percentage of Europeans against the war was much higher than the percentage of Americans against the war. In the same sort of way, we may well find some approaches to philosophy of science more strongly supported in Europe than America, while still having supporters in the USA, and conversely.

What then are the approaches to philosophy of science which are likely to find greater support in Europe? Here we are entering the realm of speculation, but I find the two tendencies mentioned by Galavotti plausible candidates. The first of these is an inter-disciplinary approach based on dialogue between scientists and philosophers. The second is the historical approach which lays emphasis on the development of scientific and philosophical concepts. After all 'old Europe' is so filled with monuments of the past that it is hard here to forget history altogether. Conversely the strong tradition of analytic philosophy in the USA may lead there to more of an emphasis on the use of formal methods and logico-linguistic analysis.

Notes

1. Any page references in the introduction which have no further specification are to the present volume.

Acknowledgements

The bulk of the funding both for the conference and the publication of the present volume was supplied by the European Science Foundation. The British Society for the Philosophy of Science also sponsored the conference. We would like to thank both organisations for their financial help which made this academic activity possible. The conference took place at King's College London and was organised through the King's Centre for Philosophical Studies. I would like to thank my college and its institutions for providing us with excellent facilities.

BIBLIOGRAPHY

[Keynes, 1936] J. M. Keynes. *The General Theory of Employment, Interest and Money.* London: Macmillan, Cambridge University Press for the Royal Economic Society, 1936. Paperback edition. 1993.
[Keynes, 1937] J. M. Keynes. The General Theory of Employment. In *The Collected Writings of John Maynard Keynes. Vol 14.* pp. 109–123. London: Macmillan, Cambridge University Press for the Royal Economic Society, 1973. (Originally published in *The Quarterly Journal of Economics*, February 1937.)

Donald Gillies

Department of Philosophy
King's College London
Strand, London WC2R 2LS, UK.
Email: donald.gillies@kcl.ac.uk

Knowledge Gain and Practical Use: Models in Pure and Applied Research

MARTIN CARRIER

1 Roles of Models in Science

Heavy emphasis has been placed on models in the philosophy of science during the past decades. This feature is in part due to the fact that quite different entities are subsumed under the heading 'model'. Consequently, universal appeal to models should not be taken to indicate that a single, uniform characteristic of science is stressed unanimously. Talk about models may refer to distinct objects such as scale models, visual representations, numerical solutions of equations, simulations, idealized or tentative hypotheses, the domains of application of theories and many more ([Stöckler, 1994, pp. 46–53]; [Hartmann, 1999, pp. 328–329]).

I want to focus on models as a means for bringing theory to bear on experience. In the most narrow understanding, models specify boundary and initial conditions that are requisite for employing laws for the purpose of explaining or predicting phenomena. For instance, the laws of celestial mechanics taken in themselves entail nothing specific about celestial orbits. Rather, in order to arrive at claims about, say, the solar system, boundary and initial conditions need to be adduced. They express assumptions about the composition of the solar system, that is, the mass ratios involved, the positions of the planets at a particular time, and so on. Only if laws are supplemented with conditions that address particular circumstances, can these laws explain and predict specific phenomena. The laws of celestial mechanics together with these factual assumptions constitute a model of the solar system. In general, models include laws of nature along with boundary and initial conditions. Models are distinguished by this inclusion of factual elements which make them suitable for spelling out the consequences of some corpus of laws for some type of situations. This means that models are conceptually distinguished by their local nature: they are restricted to a particular class of facts.

On the other hand, this view suggests that the nomological elements of models all derive from some overarching theory or theories. Nancy Cartwright was the first to point out systematically, that this is not true. As she argues, it is extremely rare in science that the models of a class of phenomena are shaped and dominated by a high-brow theory. Rather, such 'models of a theory' are most often descrip-

Laws and Models in Science, 1–17.

tively inaccurate. Cartwright's point is that these flaws are not fixed by adding theory-induced corrections or other details whose consideration is suggested by the theory. Instead, the empirical performance of a model is improved, usually, by bringing to bear phenomenological laws or practical approximations. The explanatory burden in science is borne by theory-independent auxiliaries, approximations and corrections ([Cartwright, 1983, Chap. 2–3, 6, 8], [Cartwright, 1997; Cartwright, 1998]; see [Ramsey, 1997]).

Margaret Morrison agrees with Cartwright in that the concrete models rather than the abstract theory represent and explain the behaviour of physical systems, but she shows, on the other hand, that Cartwright overstates her case. Morrison convincingly argues that high-level theories do play a significant role in structuring the phenomena and in suggesting ways to account for them. Theoretical principles form an indispensable part of the pertinent models and they bear a part of the explanatory burden for this reason. Moreover, phenomenological models are not, in general, descriptively more accurate than theoretical ones. But in accordance with Cartwright, Morrison also stresses that models invoke additional generalizations without theoretical backing, and, of course, information about particular facts. In sum, models essentially depend on theory—but only in part ([Morrison, 1998, p. 70], [Morrison, 1999, p. 39, pp. 61–63]; see [Winsberg, 2003, p. 106]).

I follow Morrison's account in assuming that models rely on theoretical principles, other nomological assumptions and claims about matters of fact to a varying degree. Since many models thrive on these extra-theoretical elements, such models are not 'models of a theory'. They cannot be constructed on the basis of an overarching theory. The necessary auxiliary assumptions may derive from another theory or from the particulars of the situation at hand, and they serve to bridge the rift between the theoretical principles and the observations. In this sense, models are supposed to 'mediate' between theory and experience [Morrison and Morgan, 1999, p. 10, 18].

Morrison distinguishes between two procedures of model building. Constructing a model typically starts with an idealized account of a phenomenon. A simplified version of the situation is considered at first in which perturbations and more complex aspects are ignored. Only the gross features are represented in the model. The ensuing process of 'de-idealization' feeds the more subtle details back into the account. In principle, there are two paths this elaboration can take. One can specify theory-based corrections which take care of all the distortions present under the conditions at hand. This process may yield a convergent series of models so that an unambiguous picture emerges. This picture contains the unique theoretical representation of the phenomenon. Morrison's example concerns pendulum motion in classical mechanics. In this case, the theory serves to highlight possible distortions and supplies the means for their correction. The series of correction factors converges toward one coherent account of pendulum motion. That is, the accumulation of these theory-induced refinements gives an increasingly more adequate picture of the phenomenon.

In other cases, by contrast, no such coherent account emerges. Rather, contradictory models prevail. Morrison's chief example concerns the atomic nucleus. The so-called liquid drop model treats nucleons analogously to particles in a liquid drop: they move rapidly and undergo frequent collisions. This model is able to account for the absolute values of nuclear binding energy and its approximate dependence on nuclear mass. The shell model takes into consideration that nucleons possess quantum properties and, in particular, obey Pauli's exclusion principle. This model takes care of nuclear spin, and adds small-scale corrections to the nuclear binding energy as estimated on the basis of the liquid drop model. The crucial aspect is that these quantum features cannot simply be added to the liquid drop approach. No single, coherent picture of the atomic nucleus emerges; we are left with incompatible models for different purposes. Examples like this suggest the interpretation that models partially represent a class of phenomena. They render only some of the relevant aspects and cannot be filled in so as to furnish a comprehensive portrait of these phenomena ([Morrison, 1998, p. 68, pp. 74–75], [Morrison, 1999, pp. 48–52]; see [Cartwright, 1983, p. 104]).

2 Methodological Features of Applied Science

The general message of the just-sketched account of models is that the particular and the factual is no less important in science than the nomological or the universal. Models incorporate the more concrete aspects of a situation and thereby serve to connect theoretical principles with observations. This account was developed for capturing methodological features of epistemically driven science. However, applied science is more at the focus of attention in these days. A large amount of research is done and sponsored because some technological innovation is sought. What matters for science is practical success: the control of natural phenomena and the intervention in the course of nature.

Applied science is of a diverse nature; it comes in varieties of different methodological orientation and different degrees of ingenuity. For instance, applied science may proceed by deduction. In such cases, the applied scientist operates like an engineer. He or she employs the toolkit of established principles and brings general theories to bear on technological challenges. Novel devices are created, but no insights into nature's workings gained. Another strategy widely employed in applied science is trial and error. In such cases, the basic features of the problem area aren't yet understood; exploratory experimentation is used to accomplish a preliminary understanding in the first place. Applied science of these sorts only rarely stimulates epistemic progress.

Other instances of applied science are more productive epistemically. Such applied research is directed at practical goals, to be sure, but it brings about knowledge gains or produces new insights. Such 'application-dominated research' is the epistemically fertile part of applied science. It is applied research in being driven by concrete pragmatic or technological interests and being directed at practical goals. But application-dominated research is similar to epistemic research in seeking new solutions to scientific problems. It does not achieve its goals by straightforward deduction from established principles, nor by using trial-and-error procedures for groping in the dark. It rather brings forth something new, a novel piece of knowledge. Application-dominated research operates at an intermediate level between applied science *simpliciter* and epistemic research.

Applied science in general is characterized by its pragmatic attitude and by its commitment to the proper functioning of some device as its chief criterion of success. In addition, applied science is put under heavy pressure from economic companies or political institutions alike to supply quickly solutions to practical problems. Science is the first institution called upon if advice in practical matters is needed. The question is what the emphasis on the search for control of natural phenomena does to science and whether it interferes with the search for knowledge.

On the face of it, there are contrasting observations and arguments. One impression is that applied science is *methodologically deficient*. The physicist Sylvan Schweber observes that 'the demand for relevance ... can easily become a source of corruption of the scientific process' and stresses the special role of 'scientists engaged in fundamental physics' in that their community is committed to the vision of truth (Schweber [1993, p. 40]). There are two influences on applied science that could make this concern of methodological deficiency plausible, namely, the probable prevalence of a purely pragmatic attitude among applied scientists and the overburdening of applied science by complexity.

First, as to pragmatism, it seems that if a gadget works, everything is fine; no further questions will be asked. Applied science appears to be governed by a pragmatic attitude which becomes manifest in a restricted scope of theorizing and explaining. Epistemic challenges that transcend immediate practical needs might be ignored. Second, as to complexity, applied science is characterized by a specific, intrinsic penchant toward the complex. In virtue of its commitment to experience, science in general needs to address empirical instances, to be sure, but not particularly intricate phenomena. On the contrary, empirical tests often proceed better by focusing on the pure cases, the idealized ones, because such cases typically yield a more direct access to the processes considered fundamental by the theory at hand. But applied science is denied the privilege of epistemic research to select its problems according to their tractability; rather, its research agenda is set from outside. Practical challenges typically involve a more intricate intertwinement of factors and are thus harder to put under control. Applied science cannot help but confront complexity.

It appears plausible that scientists respond to such excess demands by adopting *tentative epistemic strategies*. Likewise, the dominance of purely practical success criteria tends to cut off research from any deeper epistemic interests. As a result of both possible trends, applied science might resort to purely local solutions without theoretical integration. Such *local models* might be restricted to particular sets of circumstances and invoke ceteris-paribus laws, generalizations from different theoretical sources or generalizations without theoretical foundation. That is, models invoked in applied science could be at once internally heterogeneous and differ substantially from one another. Moreover, the causal claims entertained might be contextualized in that they only hold under 'normal' conditions and leave the pertinent causal processes out of consideration. Pragmatic investigations plausibly focus on such cause-effect relations that prevail under 'typical' conditions and are thus usually sufficient for bringing about the effect. Likewise, applied science might invoke pragmatic simulation techniques which pursue instrumentalist modelling strategies and deliberately leave out what really happens in the system in question. Knowledge gained in applied research contexts might look like a patchwork of local and instrumentalist models, isolated from one another and lacking deeper explanatory import.

But a contrasting train of thought sounds also plausible. The just-given argument suggests that applied science is tempted by its practical nature to adopt tentative epistemic strategies and to abandon the goal of understanding. But one might respond that the potential for intervention is linked with theoretical understanding. This intuition of *sustained epistemic quality control* is based on the assumption that superficial relations of the sort sketched will eventually fail to underwrite technological progress. Conversely speaking, theoretically understood causal relations provide much more opportunities for intervention than contextualized causal relations do. The former can more easily be generalized and applied to a wider range of conditions; they are thus more useful in a practical respect. The theoretical explanation or integration improves the prospect of bringing other factors to bear on the process at hand and to twist the latter so that it delivers more efficiently or more reliably what is demanded.

Underlying this alternative account is the assumption that control is achieved best by bringing to bear methodological standards that also characterize understanding, namely, unified explanation and causal analysis ([Kitcher, 1981]; [Salmon, 1984, pp. 135–157, pp. 206–238]). On the one hand, these virtues codify what knowledge or understanding is all about. We understand a phenomenon when we are able to embed it in a nomological framework, and we grasp a causal relation when we can take account of the process leading from the cause to the effect. On the other hand, such methodological virtues are also crucial for making sustained technological progress possible.

On the face of it, there is a point to both arguments. If the proper functioning of a device is the only achievement that counts, it can hardly be expected that scientists immerse themselves in theoretical work in order to give a sound explanation of the operation of this device. Right? But if reliability of control and trustworthiness of intervention can only be maintained and ascertained by accounting for the mechanism underlying the proper functioning of the device, there is reason to assume that applied scientists would take pains to understand this mechanism theoretically. Right? So, this is one of the philosophical issues that cannot be resolved by in-principle considerations. Rather, the methodology of applied science needs to be studied by addressing concrete cases.

3   The Role of Local Models in Application-dominated Research

I mentioned a number of tentative epistemic strategies which might be invoked in application-dominated research: local modelling, pragmatic simulation techniques, exploratory experimentation, contextualized causal relations. Here I wish to focus on the role of local models.[1] The question is whether the models invoked in applied research merely contain specific assumptions of narrow scope, only loosely tied together by shared principles. In other words, I explore whether the conceptual and nomological heterogeneity of applied research significantly exceeds that of pure or epistemic science.

The first thing to be noted in this connection is that the emphasis specifically lies on research, not on applied science in general. A large number of relevant challenges demand the development of technology rather than scientific innovation. This means that extant knowledge is brought to bear on some issue, and the operation of an invention can be understood by appeal to knowledge available beforehand. A known feature of nature is used for producing a technological feat. The development of optical switches is a case in point. Light signals are used for modifying the index of refraction of some suitable medium. This change gives rise to interference phenomena which are then employed for an extremely fast opening or closing of optical gates without intermediate electron flow (Linsmeier [2001]). This is certainly a creative advancement in technology, but the innovation only lies in forging novel ties among elements of extant knowledge. No new effect was discovered, no new theory was formulated. But I wish to focus on the structure of knowledge gained in applied research—in contradistinction to theoretical knowledge that is merely brought to bear on practical questions.

Correspondingly, it does not militate against the methodological dignity of applied science that the models invoked in developing or designing apparatus are fragmentary or known to be false. For instance, models from geometrical optics

---

[1] See [Carrier, forthcoming] for the problem of contextualized causal relations.

are widely used for designing optical instruments such as lasers. The path of light is calculated on the basis of a superseded theory that leaves out known facts about the physical nature of light. But this purely technological use implies nothing concerning the methodology of applied research. Such models lack epistemic impact: no new factual claim is made, no new hypothesis is generated.

Turning now to application-dominated research proper, the worries about methodological deficiency are found confirmed to some extent. I briefly sketch three examples. The first one is taken from nanoresearch and features the disregard of welcome anomalies. One of the top items on the nanoscale research agenda is the attempt to build molecular wires: the current should be conducted through individual, ring-shaped organic molecules. After the effect of molecular conduction was established, researchers were satisfied. In particular, they failed to inquire into a glaring discrepancy that had shown up along the way, namely, that the resistance was much lower than anticipated. This unexpected finding increased the options for practical use. No systematic attempts were made at clarifying the cause of this welcome anomaly. An ad-hoc-hypothesis without theoretical or empirical backing was offered and the case was closed with emphasizing the technological prospects offered by molecular conduction ([Nordmann, 2003]; see [Reed and Tour, 2000, pp. 90–91]). This failure to address anomalies and the missing attempts to elucidate the mechanisms underlying a practically promising effect confirm the suspicion of methodological deficiency. The measured value of molecular conduction was simply accepted as the basis of all further consideration. This essential parameter was read off from experience, and hardly any serious attempt was made to link it with the system of knowledge. The anomaly was glossed over by a purely local adaptation. The narrow scope of theorizing and explaining manifest in this example testifies that deeper epistemic interests were not pursued.

The second example concerns 'giant magnetoresistance', a physical effect discovered in 1988 and quickly explored by industrial research laboratories. This effect involves spin-dependent scattering of electrons and makes it possible to build extremely sensitive magnetic field sensors. Giant magnetoresistance underlies the functioning of today's magnetic read heads; it is used for hard disks or magnetic tapes. The qualitative explanation of the effect was suggested immediately after its discovery, but building a suitable device requires knowledge of quantitative relations. The avowed aim of the industrial research laboratories involved in this research was to come up with what is sometimes called 'design rules'. Such experimentally confirmed rules provide approximate relations among relevant parameters (such as layer thickness or ferromagnetic coupling between layers) and can thus be employed for manufacturing read heads. Design rules of this sort are extremely local. They apply to particular types of read heads only—with the result that the design rules used by Philips don't work for IBM read heads. One realizes that design rules are part of models of a fairly restricted scope [Wilholt, 2003].

The third example is taken from biotechnology; it does not concern local models but contextualized causal relations [Carrier, forthcoming]. Still, this case likewise exhibits the restricted scope of theorizing in application-dominated research and thus underlines the importance of local approaches in applied research. The background of this case is the present turn in cell biology from genomics to proteomics. In contrast to earlier views about rigid connections between properties of a cell and its genetic makeup, the interaction among proteins is now increasingly stressed. While the conceptual and theoretical insufficiency of genomics is widely acknowledged, biotechnologists tend to hold fast to invariant gene-property relations in order to retain their 'handle' for intervention. Stimulation of a certain gene is usually sufficient for producing some cell property. That is, genetic manipulation is suitable for bringing about effects in a predictable fashion, albeit the reliability of the relevant causal relations is constrained to particular conditions [Keller, 2000, pp. 141–142]. Arguments from the biotechnological camp concede that genetic determinism is superseded and discarded in bioscience, but insist that it rules unquestioned in biotechnology. The assumption of a close connection between gene and cell property constitutes a lever for intervening in biological processes. Genes are tools for bringing about intended effects and for achieving biotechnological progress. Scientific truths are said to be unnecessary for this purpose. Technology aims at practical success which is accomplished by the identification of levers to pull and switches to press [Bains, 1997].

The upshot of all three cases is that models advanced in applied research are sometimes of a fairly narrow scope and merely apply to a small set of circumstances. As a result, such models fail to provide deeper knowledge but only support a specific technological use. Scientific understanding and capacity of intervention appear to be decoupled. In a similar vein, Helen Longino argues that in a context like industrial manufacture the important goal is to build a device that operates in a desired way (such as a cell system for producing human insulin). In such a context, she grants, understanding may not be necessary. Longino takes this assumed variability of methodological judgement as a confirmation of her pluralist approach to science [Longino, 2002, pp. 200–201].

The general impact of these considerations is that applied scientists might appraise theoretical proposals exclusively on their potential for intervention. It would be sufficient for accepting such proposals that they enable control. Achieving a technological goal is of exclusive relevance; gaining understanding of nature is an epistemic luxury that applied research cannot afford. Such methodological features of applied science nourish the suspicion that science suffers from the grip of practical demands. Theories of epistemic science are expected to excel in virtues like explanatory power, predictive force or unifying capacity. By contrast, applied science seems to be characterized by local models, superficial generalizations and theoretical heterogeneity. The explanatory and unifying bearing of theory apparently fails to extend to technological challenges. On the face of it, applied research is methodologically inferior to epistemic science.

However, this conclusion appears to be premature and incompatible with the thrust of section 1 on the roles of models in science. After all, there is a striking contrast between the views of Cartwright or Morrison on extra-theoretical nomological ingredients of models, on the one hand, and the just-rehearsed commitment of epistemic science to unifying explanation, on the other. Models are used in epistemic and applied science alike, and it is plain from Morrison's (or Cartwright's) analysis that the former are less coherent and unifying than one might have thought. For instance, Cartwright suggests that a general tension obtains between accuracy and unification. The more precise the results are required to be, the more dappled the conceptual resources have to be [Cartwright, 1983, pp. 104–107]. In other words, fundamental science likewise appears to employ epistemic strategies that might be considered tentative. So, let's turn to more advanced cases and see if significant parallels between epistemic and applicative reasoning turn up.

4 Model-building Strategies in Epistemic Research

One of the lessons to be drawn from Morrison's analysis of model-building strategies in epistemic research is that the nomological parts of models can only rarely be derived in their entirety from overarching theory. [2] Such a theory typically leaves a great deal of interstices to be filled by drawing on other nomological resources. This lesson brings a changed understanding of pure or epistemic science in its train. Namely, it is difficult for higher-order theories in general to reach the level of concrete experience. And this is why such theories have a hard time guiding or directing applied research. It is true, applied science has to tackle the demand to address particularly intricate situations (see Section 2), but the overriding tendency is that universal principles don't excel in capturing the subtle details of experience. This is a predicament of science as a whole. In epistemic science it becomes manifest as the difficulty of linking up theory with fact, in applied science it materializes as the difficulty of translating general insights into working devices. Unaided universal principles often fail to extend to multi-faceted and complex experience. This is why explanations both in epistemic and applied science sometimes fall short of the ideal of unification; and this is why model-building needs to resort to strategies that have the air of being tentative.

These considerations suggest the following twofold claim. First, the pragmatic attitude as it prevails in applied science indeed contributes to lowering the methodological demands placed on models that are considered acceptable. This claim is

---

[2] Actually, it is a widely shared insight today that having accepted overarching theoretical principles at hand still leaves a lot of challenges for building models. This insight could also be attributed to Ian Hacking, Cartwright and Ron Giere (see [Winsberg, 2003, p. 120]).

confirmed by the examples given in Section 3. But, second, this reduction is less marked than anticipated. That is, model-building strategies in applied research are methodologically deficient—but to a smaller extent than anticipated in light of Section 3. Let me support this second contention by presenting relevant examples. I want to show two things. On the one hand, epistemic science also resorts to model-building strategies that look tentative and comes up with models that appear local and heterogeneous. On the other hand, local models may constitute an epistemic virtue rather than a vice. As the case may be, local models may even promote understanding.

The construction of a fusion reactor is among the challenges of technology development. One of the relevant attempts is the so-called Tokamak reactor in which hydrogen plasma is confined by a suitably structured magnetic field. The treatment of the torus-shaped plasma proceeds by dividing the relevant realm into two different sectors, the core and the exterior, characterized by open or closed magnetic field surfaces, respectively. In fact, a continual transition between these two regimes obtains. But the theoretical treatment introduces a sharp boundary and thus creates two distinct sectors. This procedure yields two different models that apply, respectively, to these sectors. In the resulting overall account, these disparate models are simply joined, they are stitched together [Stöltzner, 2003].

Piecing together disparate models certainly looks like a pragmatically driven and epistemically insufficient strategy. What should have been considered a homogeneous material with gradual transitions is divided into separate boxes with discrete property changes at the boundary. Consequently, this boundary is deliberately artificial so that the realist commitment underlying epistemic research is relinquished in favour of attaining a practical goal, namely, designing a mechanism of plasma confinement. It seems that application dominated research indeed suffers from methodological deficiency.

In fact the example does not support the conclusion. In 1904, Ludwig Prandtl used a similar procedure in order to give the first empirically adequate account of the flow of a viscous fluid. Such processes are governed by the Navier-Stokes equation which, however, is unsolvable in general. Prandtl's innovation was a conceptual distinction between two regions of fluid flow: a thin layer near the surface of bodies (such as the walls of a water pipe or objects in the pipe) and the remainder. In the first region the fluid adheres to the surface; no slip occurs. The viscosity is assumed to dominate the flow. In the second region, the viscosity is supposed to be negligible. Both assumptions are known to be incorrect. Water possesses a low viscosity which is, consequently, neither high, as in the first case, nor zero, as in the second. This means, in sum, that Prandtl invoked idealizations; in fact, he devised two idealized cases separated by a sharp boundary. These two cases could be treated mathematically; the Navier-Stokes equation becomes solvable under such simplifying constraints [Morrison, 1999, p. 46, pp. 53–61].

The cases of the fusion reactor and the water pipe exhibit a similar methodological pattern: two disparate models are stitched together [Stöltzner, 2003].[3] A uniform variation is replaced by a sharp boundary; a unifying treatment is supplanted by two local models. But in the hydrodynamics case this pattern cannot be attributed to application dominance. Prandtl's work was epistemic research; it was intended to resolve some glaring anomalies of hydrodynamic theory.

Local models play an important role in epistemic and applied science alike. This methodological parallel is underlined more strongly by the similarity of the just-presented cases with the earlier example of using two conflicting models of the atomic nucleus (see Section 1). The liquid drop model and the shell model are incompatible but they serve to account for different properties of the nucleus. Similarly, in the cases of fusion and fluid flow, the two disparate models are applied to distinct regions. All these examples converge in showing that likewise in epistemic research (on nuclear properties or fluid flow) and in applied research (on the fusion reactor), heterogeneous and conflicting models are employed in order to arrive at a comprehensive explanation of the phenomena.

A possible response is that methodological deficiency is more widespread than suspected initially and extends to parts of epistemic research as well. Tentative strategies impair epistemic research, too. But such a complaint would go astray. One of the presuppositions of the preceding discussion was that local modelling is methodologically suspect, that it is detrimental to epistemic aspirations and produced by strong application pressure. The assumption was that understanding is generated by overarching theories, while local modelling indicates a dominance of pragmatic interests. But this is not true—at least not universally. Local modelling is not, in general, methodologically inferior. Instead, such models may be necessary to give an empirically adequate account of the phenomena and they may create understanding rather than detract from it.

First, as to empirical adequacy. Explaining and predicting the empirical properties of a phenomenon is no doubt among the goals of epistemic science. But to succeed in this endeavour under complex circumstances may require the inclusion of particular conditions and considerations. Complaints about the specificity or heterogeneity of the knowledge sources drawn upon are less than convincing if the alternative is empirical failure. Taking the particulars into account may be the only chance for coming to terms with the relevant phenomena in the first place.

Consider a contrary case as an illustration. Hydrogeological models tend to operate with idealizations of the water currents and soil conditions. In applying such models to the analysis of groundwater flow, these limitations are disregarded

---

[3]The numerical treatment of shock discontinuities exhibits the same feature: it involves piecing together several partial solutions [Winsberg, 2003, pp. 122–123]. Accounting for shock waves is of prime importance in several practical areas, but it constitutes a theoretical challenge as well. Again, the seemingly tentative procedure of stitching together local accounts is found in epistemically relevant research as well.

with the result that the inferences drawn are unreliable and frequently off the mark [Shrader-Frechette, 1997, pp. 153–154]. In this case the trouble does not arise from indulging in details and severing all ties to high-brow theory. Quite the contrary. The problem rather lies in the overreliance on the general theoretical approach and the vain attempt to derive the models from this framework. These models fail because they are insufficiently local. In order to capture the phenomena, the unique, local conditions need to be integrated in the model [Beven, 2001, pp. 4–6]. This strikes a Cartwrightian chord: in empirical respect going local is often an asset rather than a liability (see Section 1). And empirical adequacy is certainly a chief commitment of epistemic science.

Second, as to understanding produced by local models, consider the hydrodynamics example. The pertinent overarching law is the Navier-Stokes equation. But even in cases in which this equation is solvable, it does not provide a mechanism or an intuitive access to the observed effects. The calculation is like a black box; it yields numbers but no understanding. Similar observations have been made with respect to Schrödinger's equation or the standard model of particle physics ([Hartmann, 1999, p. 329]; [Batterman, 2000, p. 232]). By contrast, Prandtl's local, unrealistic models provide a mechanism and some sort of causal understanding of the processes involved in fluid flow.

The explanatory power of local models can also be identified in other cases. The laws of quantum chromodynamics and the models of the pertinent nucleons exhibit the same relation. Quantum chromodynamics is the fundamental theory of nuclear interaction, yet it fails to provide an intuitive understanding of the relevant processes and mechanisms. In response to this defect, models are conceived that are inspired by the formalism of the theory, but cannot be deduced from it. Moreover, different such models are employed for explaining different properties. It is these models that specify the relevant mechanisms and outline causal processes at work [Hartmann, 1999, pp. 331–344].

The explanatory impact of local models can be recognized in other contexts as well. As Robert Batterman argued, understanding may be reached by drawing on idealized, simplified and approximate models. In contradistinction, taking the details unrestrictedly into account and giving a full-scale derivation of the particulars of the relevant phenomena from first principles may obscure their salient features. One of Batterman's examples concerns phase transitions. Models relying on relations between macroscopic quantities—what Batterman calls 'minimal models'—can be employed for equally describing the behaviour of fluids and of ferromagnetic substances in the vicinity of what is called the critical point. Quantities like pressure and temperature or magnetization and temperature, respectively, are connected by the same straightforward relation. This account disregards the distinct microstructures underlying the two kinds of phenomena. The pertinent models are local in that they explicitly ignore the universal physical principles that are acknowledged to govern the phenomena in question and instead draw on generalizations that, due to the idealizations involved, do not hold true, strictly speaking, of these phenomena.

Minimal models reveal that the behaviour of physical systems as diverse as fluids and magnets is identical if viewed from a particular angle. The invariant pattern characteristic of phase transitions is hard to extract from an analysis of the respective microstructures. The reason is that the microphysical details are distinct in each case and thus fail to explain the emergence of a shared property. Such challenges are best met by introducing quantities that do not reflect the underlying microphysical differences. Minimal models provide simplified and idealized accounts and they are able for this reason to highlight the common ground among the diverse manifestations of a physical effect [Batterman, 2002, pp. 22–27].

The upshot is that local models are capable of playing a constructive epistemic role. They may provide the only route to an empirically adequate account and they may create understanding where comprehensive theories may yield nothing but opaque calculations. I tried to outline two such local pathways to understanding. First, a web of more restricted accounts may offer intuitive access to the underlying causal chains while general theories may merely produce numerically adequate outcome. Second, Batterman's minimal models provide understanding by unifying phenomena that appeared disparate otherwise. Phenomena that are different in appearance and with regard to their microphysical structure—such as evaporation and the loss of ferromagnetism during temperature increase—are shown to instantiate the same macroscopic relation. Minimal models forge links between phenomena that are dissimilar as judged by their microconstituents and microprocesses. These achievements suggest that local modelling should not be taken as an epistemic failure across the board.

On the other hand, this case must not be overstated. Comprehensive theories also engender understanding—if of a different sort. In virtue of their generality, they connect a vast number of phenomena and establish similarity relations among huge classes of phenomena. Local models, in view of their local nature, fall short of this standard. This holds true even of Batterman's minimal models since the set of phenomena related by any such model is much smaller than, say, the class of phenomena accounted for by the theoretical models of classical mechanics. Local models thus fail, in general, to provide understanding in the unificatory sense. But they frequently specify causal mechanisms. Roughly speaking, comprehensive theories produce understanding by unification, local models may contribute to understanding by supplying abstract laws with causal mechanisms. So, overarching theories and local models may both generate understanding—each in their own way.

Note, however, that in no way do all local models provide understanding. I gave a number of examples to the contrary effect (see Section 3). Local models may encode nothing but superficial relations between insignificant and theoretically barren quantities. They may be read off from the data and restricted to a narrow range of parameters. My intention was to bring out the explanatory *potential* of local models and envisage what they can achieve when they are at their best.

The overall message is that epistemic and applied research resemble one another more strongly than anticipated. While it is true that we find tentative epistemic strategies and an increased level of superficiality in applied research, we also encounter research strategies with a clear epistemic bearing. Applied research, too, is at least in part committed to virtues like unification and causal analysis. Whereas these methodological criteria may have a traditional ring, they manifest themselves in a less customary way. These methodological goals are pursued not by articulating comprehensive nomological principles but rather by building a web of interlinked and intertwined local models. Unification and causal analysis is not implemented by erecting a hierarchy of laws of nature, but by piecing together a network of models, each restricted in scope but hooked up with many others by shared assumptions.

5 The Question Dynamics of Applied Research

My considerations proceeded from the following set of suppositions: First, applied research is characterized by a preference for local models; second, this preference is an indication of methodological deficiency; third, this deficiency is the result of a prevalence of purely practical concerns and of the overtaxing of applied science by complexity. The preceding line of argumentation tends to undermine these initial assumptions. Local modelling is not a distinct feature of applied research; local models rather play a significant role in epistemic research as well. And they do so with good reason since they can be instrumental in fostering understanding. It is true, some of the worries concerning the methodological inferiority of applied research are justified. The nanoconduction example shows that glaring anomalies are glossed over if they don't interfere with practical use. The magnetoresistance case gives testimony to the inclination to be satisfied with superficial relations of immediate bearing on technical problems. The genetics example bears witness to the willingness to adopt approaches known to be false if they appear suitable for advancing technology (see Section 3). Science may suffer from application pressure in methodological respect.

On the other hand, there is a methodological protection built into applied research that tends to uphold demanding criteria of acceptance and thus to keep the epistemic loss at a moderate level. Epistemic and applied research share important methodological characteristics. And this is no coincidence but rather brings out a distinctive feature of science: understanding tends to improve the options for intervention. This has, no doubt, a familiar ring: knowledge is power. This result lends credibility to the second train of thought above, intended to elaborate the intuition of sustained epistemic quality control (see Section 2). Both unified explanation and causal analysis codify what understanding is all about. We understand a phenomenon when we are able to embed it in a nomological framework, and we grasp a causal relation when we can account for the process leading from the cause to the effect. These same virtues are also essential for successful research on practical matters. Unification forges links among phenomena and makes it possible to apply some result within a large domain. Causal analysis elucidates the steps leading up from the antecedent to the final state and thereby offers opportunities for technological control. The effect can be modified and adapted by intervening at one of the intermediate stages.

At a second glance, the above-given examples for the methodological deficiency of local models (see Section 3) tend support this view. First, one of the research groups operating in the field of nanoconduction explicitly aims to hook up the observed effects with fundamental theory [Nordmann, 2003]. Second, the 'design rules' sought for bringing to bear giant magnetoresistance on technology development are not, in fact, established by induction. They do not originate from some theoretically uninformed screening procedure, but they are rather theoretically derived and experimentally examined [Wilholt, 2003]. Overarching theories continue to play an important role in dealing with practical challenges. Third, in contrast to the allegations of some biotechnologists, the substitution of genomics by proteomics in bioscience had important ramifications on biotechnology. For instance, one of the foci of recent biomedical research is represented by the question which genes are actually switched on; that is, which genes produce proteins. Questions of this sort transcend the genomic horizon and indicate that biotechnology has entered the proteomics age [Carrier, forthcoming, Section 7].

Accordingly, science is faced with a question dynamics leading from applied issues to fundamental ones. For methodological reasons, applied research tends to transcend practical questions and grows into epistemic research. Practical challenges often bring fundamental problems in their train. Such challenges cannot appropriately be met without treating these fundamental problems as well. Understanding is not among the explicit objectives of applied science, but once in a while it still produces epistemically significant insights. This feature I call *application innovation*. It involves the emergence of theoretically significant novelties within the framework of use-oriented research projects. For instance, conceptually revolutionary innovations in biology such as 'retrovirus' or 'prion' were introduced within the practical research context of identifying infectious chains. It follows that the primacy of application need not pose a threat to the epistemic dignity of science. Not infrequently, epistemic insight is the unintended by-product of addressing practical needs thoroughly. Theoretical unification and causal analysis are inherent in both pure and applied research, because such methodological virtues promote understanding and intervention at the same time.

BIBLIOGRAPHY

[Bains, 1997] W. Bains. Should We Hire an Epistemologist?, *Nature Biotechnology*, **15**, 396, 1997.

[Batterman, 2000] R.W. Batterman. A 'Modern' (=Victorian?) Attitude Towards Scientific Understanding. *The Monist*, **83**, 228–257, 2000.

[Batterman, 2002] R.W. Batterman. Asymptotics and the Role of Minimal Models. *The British Journal for the Philosophy of Science*, **53**, 21–38, 2002.

[Beven, 2001] K. Beven. How far can we go in distributed hydrological modeling? *Hydrology and Earth System Sciences*, **5**, 1–12, 2001.

[Carrier, forthcoming] M. Carrier. Knowledge and Control: On the Bearing of Epistemic Values in Applied Science. In P. Machamer and G. Wolters (eds.), *Science, Values, and Objectivity*, Pittsburgh: University of Pittsburgh Press, forthcoming.

[Cartwright, 1983] N. Cartwright. *How the Laws of Physics Lie*. Oxford: Clarendon Press, 1983.

[Cartwright, 1997] N. Cartwright. Models: The Blueprints for Laws. *PSA 1996 II. Philosophy of Science Supplement to Volume 64*, 292–303, 1997.

[Cartwright, 1998] N. Cartwright. How Theories Relate: Takeovers or Partnerships? *Philosophia Naturalis*, **35**, B. Falkenburg and W. Muschik (eds.), 23–34, 1998.

[Keller, 2000] E. Fox Keller. *The Century of the Gene*. Cambridge Mass.: Harvard University Press, 2000.

[Hartmann, 1999] S. Hartmann. Models and Stories in Hadron Physics. In [Morgan and Morrison, 1999, pp. 326–346].

[Kitcher, 1981] P. Kitcher. Explanatory Unification. In: J.C. Pitt (ed.), *Theories of Explanation*, pp. 167–187. New York: Oxford University Press, 1988.

[Linsmeier, 2001] K.-D. Linsmeier. Lichtschalter für Glasfasernetze. *Spektrum der Wissenschaft*, 2/2002, 76–78, 2001.

[Longino, 2002] H.E. Longino. *The Fate of Knowledge*. Princeton: Princeton University Press, 2002.

[Morgan and Morrison, 1999] M.S. Morgan and M. Morrison (eds.). *Models as Mediators. Perspectives on Natural and Social Sciences*. Cambridge: Cambridge University Press, 1999.

[Morrison, 1998] M. Morrison. Modelling Nature: Between Physics and the Physical World. *Philosophia Naturalis*, **35**, B. Falkenburg and W. Muschik (eds.), 65–85, 1998.

[Morrison, 1999] M. Morrison. Models as Autonomous Agents. In [Morgan and Morrison, 1999, pp. 38–65].

[Morrison and Morgan, 1999] M. Morrison and M.S. Morgan. Models as Mediating Instruments. In [Morgan and Morrison, 1999, pp. 10–37].

[Nordmann, 2003] A. Nordmann. Molecular Disjunctions. Lecture given at the conference Discovering the Nanoscale, Columbia, South Carolina, March 2003.

[Ramsey, 1997] J.L. Ramsey. Between the Fundamental and the Phenomonogical: The Challenge of the Semi-Empirical Methods. *Philosophy of Science*, **64**, 627–653, 1997.

[Reed and Tour, 2000] M. Reed and J. Tour. Computing with Molecules. *Scientific American*, **6**, 86–93, 2000.

[Salmon, 1984] W.C. Salmon. *Scientific Explanation and the Causal Structure of the World*. Princeton: Princeton University Press, 1984.

[Shrader-Frechette, 1997] K. Shrader-Frechette. Hydrogeology and Framing Questions Having Policy Consequences. *Philosophy of Science*, **64** *(Supplement)*, S149–S179, 1997.

[Schweber, 1993] S.S. Schweber. Physics, Community and the Crisis in Physical Theory. *Physics Today*, 34–40, 1993.

[Stöckler, 1994] M. Stöckler. Theoretische Modelle. Beispiele zum Verhältnis von Theorie, Modell und Realität in der Physik des 20. Jahrhunderts,. In: H.J. Sandkühler (ed.), *Theorien, Modelle und Tatsachen. Konzepte der Philosophie und der Wissenschaften*, pp. 45–60. Frankfurt, Peter Lang, 1994.

[Stöltzner, 2003] M. Stöltzner. Application Dominance and the Model Web of Plasma Physics. Lecture given at the conference *Models, Simulation, and the Application of Mathematics, Center for Interdisciplinary Research*, Bielefeld University, June 4–7, 2003.
[Wilholt, 2003] T. Wilholt. Design Rules: The Nature of Local Models in Industry Research on Giant Magnetoresistance. Lecture given at the conference *Models, Simulation, and the Application of Mathematics, Center for Interdisciplinary Research*, Bielefeld University, June 4–7, 2003.
[Winsberg, 2003] E. Winsberg. Simulated Experiments: Methodology for a Virtual World, *Philosophy of Science*, **70**, 105–125, 2003.

Martin Carrier

Department of Philosophy
Institute for Science and Technology Studies
Bielefeld University
P.O.B. 100 131, 33501 Bielefeld, Germany.
Email: MCarrier@philosophie.uni-bielefeld.de

Opinion Dynamics:
Insights by Radically Simplifying Models

RAINER HEGSELMANN

Introduction

Think of

- a group of people, for instance a *group of experts* for something;

- each expert has an *opinion* on the topic under discussion, for instance the probability of a certain type of accident;

- *nobody is totally sure* that he is totally right;

- to some degree everybody is *willing to revise* his opinion when informed about the opinions of others, especially the opinions of *'competent' others*;

- the revisions produce a new opinion distribution which may lead to *further revisions* of opinions, and so on and so on.

Does such an opinion dynamics stabilize? Are there typical final results? When is consensus feasible? All these questions will be addressed in the following. It will be done by means of a radically simplifying model, i.e. a model based on a very small set of assumptions which are 'easy to handle'.

In the first section I will describe a basic and minimal model. Section 2 presents simulation results and gives explanations for the most important phenomena. Section 3 gathers 10 objections against the model. Sections 4 to 11 take up all the objections and discuss extensions of the original model that try to integrate the criticisms. The final section puts together some major lessons about the radically simplifying modelling approach.

1   The Model

In the past several models of opinion dynamics were developed. [1] It started with French [1956]. Important early steps did Harary [1959], Abelson [1964], De Groot

---

[1] For a systematic and historical overview see [Hegselann and Krause, 2002, Sections 2 and 3].

Laws and Models in Science, 19–46.
© *2004, the author.*

[1974], and Chatterjee [1975; 1977]. Major attention received the book *Rational Consensus in Science and Society*, [Lehrer and Wagner, 1981]. The book presents a model in which the opinion dynamics is driven by iterated *weighted* averaging. The weights reflect the competence an individual assigns to other individuals.[2] It is shown that for lots of distribution patterns of weights the individual reach a consensus whatever their initial opinions might be. The book does not deal with the question of how to assign the weights—and exactly that will be the decisive point in the model presented *here*: the paper analyses a model in which all individuals take into account the opinions of those others whose opinions are reasonable in the sense, that their opinions are not 'too strange', i.e. *not too far away from their own opinion.*

More formally and more in detail: There is a set of n individuals $i, j \in I$. Time is discrete $t = 0, 1, 2, \ldots$. Each individual starts with a certain opinion, expressed by a real number $x_i(t_0) \in [0, 1]$. Each individual i takes into account only those individuals j for which $|x_i(t) - x_j(t)| \leq \varepsilon$, the confidence interval. Thus, the set of all individuals that i takes seriously is

$$(1) \qquad I\left(i, x(t)\right) = \{1 \leq j \leq n \mid |x_i(t) - x_j(t)| \leq \varepsilon\}.$$

Individuals *update* their opinions. The next period's opinion of individual i is the *average* opinion of all those which i takes seriously, i.e.:

$$(2) \qquad x_i(t+1) = |I\left(i, x(t)\right)|^{-1} \sum_{j \in I(i, x(t))} x_j(t).$$

Thus the very essence of the model—formally a n-dimensional *non*-linear dynamical system—is *'averaging over all opinions within one's confidence interval'*. The model is called the *bounded confidence* model (*BC*-model). The model was formulated and studied for the first time by Krause [1997; 2000].[3] A model similar in spirit has recently been investigated in [Weisbuch *et al*, 2001] and [Deffuant *et al.*, 2000]. Hegselmann and Krause [2002] combines an analytical *and* simulation based approach; the result is a very general understanding of the *BC*-model.

2 Simulations

In this section we will explore the BC-*model* by means of simulations. The simulations follow the KISS-principle: *Keep it simple, stupid!*. We will to do that, firstly, in a systematic way, and , secondly, following the KISS-principle: *Keep it simple, stupid!* Under that principle it is fairly natural to start with and symmetric ε-intervals. *Homogeneity* means that the size and shape of the confidence interval is the same for *all* agents. *Symmetry* means, that the interval has the same size to

---

[2]See [Deffuant *et al.*, 2000, Section 10].
[3]See also Dittmer [2000; 2001].

the left *and* to the right, i.e. $\varepsilon_{left} = \varepsilon_{right} = \varepsilon$. Thus, the total size of the confidence interval is 2ε. Additionally, the confidence interval is *constant over time*. Opinion space is the 1-dimensional and continuous unit interval. All agents apply an *arithmetic* mean.

Given the opinion space $[0, 1]$ only confidence intervals $0 \leq \varepsilon \leq 1$ make sense. It seems likely that the size of ε matters. Following that intuition we will *stepwise* increase the confidence interval, 'walking' from 0 direction 1, and analyse the results.

All simulations start with an uniform random distribution of 625 opinions. Updating is *simultaneous*. Figure 1 shows three stops on the tour from 0 direction 1. The stops are *single* runs. All simulations get going with the *same* start distribution. The ordinate indicates the opinions. The abscissa represents time and shows the first 15 periods. Obviously it takes less than 15 periods to get a stable pattern: With the quite small confidence interval of $\varepsilon = 0.01$ exactly 38 different opinions survive in the end. Under the much bigger confidence of $\varepsilon = 0.15$ the agents end up in *two camps*, and with $\varepsilon = 0.25$ the result of the dynamics is *consensus*. Obviously the size of the confidence interval matters.

Figure 1 shows only *single* runs. A systematic approach requires more. For a representative analysis we will stepwise increase ε, i.e. start simulations with $\varepsilon = 0; 0.01; 0.02, 0.03, \ldots 0.04.$. It will turn out that there is nothing new and interesting for $\varepsilon > 0.4$. For each of these 41 steps we repeat the simulation 50 times, always starting with a different uniform random distribution. Each run is continued until the dynamics becomes stable. Figure 2 gives an overview.

The *x-axis* in Figure 2 represents the opinion space divided into 100 intervals. Our 41 steps of increasing confidence are represented by the *y-axis*. (Note: the steps are *not* time steps!) The *z-axis* represents the average (!) relative frequencies of opinions in the 100 opinion intervals of the opinion space after the dynamics has stabilized.

Figure 2 deserves careful inspection: At the beginning of our walk along the*y*-axis of Figure 2 there is only little confidence. For example, step 2 or 3 means that $\varepsilon = 0.01$ or $\varepsilon = 0.02$. In terms of *single* runs we are speaking about an opinion dynamics like that in Figure 1a. The *z*-values show that under little confidence we find (as an average) a small fraction of the opinions in *all* intervals of the opinion space. (Note: That does *not* imply that in a *single* run all intervals are occupied after stabilisation; see Figure 1a.) *No part* of the opinion space seems to have especially high or especially low frequencies. That changes as we step further and increase the confidence interval. Look, for instance, at step 16, i.e. $\varepsilon = 0.15$. A *single* run example for a dynamics based on confidence intervals of this size is given by Figure 1b, where we end with two camps holding different opinions. Figure 2 indicates that this is a fairly typical result: As the confidence increases the average distribution of stabilised relative frequencies becomes less and less

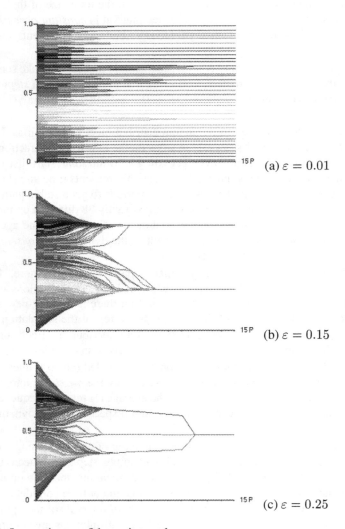

(a) $\varepsilon = 0.01$

(b) $\varepsilon = 0.15$

(c) $\varepsilon = 0.25$

Figure 1. Increasing confidence intervals

This figure is originally a coloured figure, in which colours from red over yellow, green, and blue to violet encode the opinions of the start profile. The same is the case for Figures 4, 2 and 14. Here they are all reprinted as black and white figures

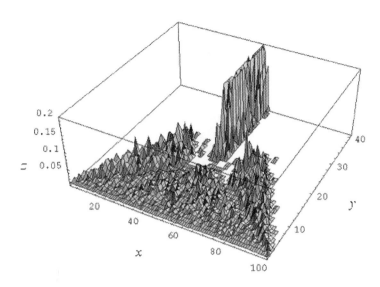

Figure 2. Walking in 41 steps from $\varepsilon = 0$ to $\varepsilon = 0.4$

uniform. To the left and to the right of the centre mountains of increasing height emerge. As we continue our walk the landscape changes dramatically again: At about step 26, i.e. $\varepsilon = 0.25$ the *'Lefty'* and the *'Righty Mountains'* come to a sudden and steep end. At the same time a new and steep centre mountain emerges. An example for a *single run* in this region of confidence is given by Figure 1c. Figure 2 shows that this single run with a total confidence interval including half of the opinion space is a typical run, leading to a consensus including all or almost all agents. Except in the centre almost all opinion intervals are empty and the corresponding opinions are eliminated.

Figure 2 shows *average* relative frequencies of opinions in the opinion space $[0, 1]$, divided into 100 intervals. Thus the average number of occupied intervals, which are actually occupied on the average is *not* directly visible *in* Figure 2. The missing information is given by Figure 3a, which shows the average numbers of occupied classes after stabilisation as we make our 41 steps (x–axis) of increasing confidence. While in Figure 2 the classes are 10^{-2} intervals of the opinion space (otherwise one gets problems with visibility), we use for the count in Figure 3a a finer scale, namely 10^{-4} intervals. Figure 3b shows the *coefficient of variation* for the values in Figure 3a

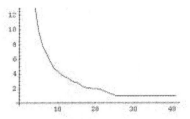

(a) Average number of occupied classes

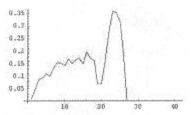

(b) Coefficient of variation for Figure 3(a)

Figure 3.

Figures 2, 3a and 3b demonstrate what Figure 1 only illustrates: Whatever the size of ε may be, under symmetry, homogeneity, constancy, and arithmetic mean we always get in *finite* time a *stable* opinion pattern.[4] Let' s refer to that phenomenon as *fragmentation*. With increasing confidence the structure of this fragmentation changes: We transit from phase to phase. More exactly, we step from *plurality* over *polarity* to *consensus*.

For an explanation and understanding of these results we will in a first step assume a *regular* start profile with $x_i(t_0) = \dfrac{i-1}{n-1}$ for all $1 \leq i \leq n, n > 1$. In such a profile the distance between two neighbouring opinions in the start profile is always the *same constant*, which makes things simpler. Figure 4 shows such a profile. Additionally we have to introduce a bit of terminology: An opinion profile $x(t) = (x_1 t), \ldots x_n(t))$ is an *ordered profile* iff $x_i(t) \leq x_{i+1}(t)$ for all $1 \leq i < n$. An ordered profile is an ε-*profile* iff $x + i+(t) - x_i(t) \leq \varepsilon$ for all $1 \leq i < n$.

Figure 4 gives decisive hints for understanding fragmentation. A grey area between two neighbouring opinions indicates that the distance between the two opinions is not greater than ε. If the area between two neighbouring opinions is *not*

---

[4]It can be shown *analytically* that the dynamics stabilizes in finite time; see[Hegselann and Krause, 2002]. The simulations give some insights about the structure of the attractor, whose existence is for sure.

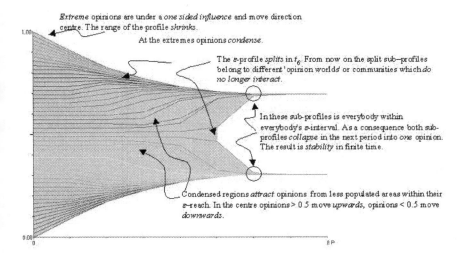

Figure 4. Shrinking, condensing, ε-split, collapse, and stability; $\varepsilon = 0.2$

grey, that indicates something we will refer to as an ε-split in the profile. Careful inspection shows:

- *Extreme* opinions are under a *one-sided influence* and move direction centre. As a consequence the range of the profile *shrinks*.

- At the extremes opinions *condense*.

- Condensed regions *attract* opinions from less populated areas within their ε-reach.

- In the centre opinions > 0.5 move *upwards*, opinions < 0.5 move *downwards*.

- The ε-profile *splits* in period t_5. From now on the split sub-profiles belong to different 'opinion worlds' or communities which *do no longer interact*.

- After the ε-split the sub profiles continue to converge. In t_6 within both sub profile everybody is in everybody's ε-interval. As a consequence both sub-profiles collapse into one and only one new opinion. From then on we have a polarization, which is stable forever.

The splits—and there may be many—do not only explain polarity, they explain the stabilisation of our dynamics in general. For fairly small confidence intervals the stabilisation leads to a fairly high number of surviving opinions, i.e. *plurality*. For 'middle sized' confidence intervals we get only a small number of surviving opinions, i.e. *polarisation*. Under large confidence intervals the profile never splits or the splits leave alone extreme and quite small minorities while an overwhelming majority converges in the centre of the opinion space. The effects described so far do not essentially depend on the regularity of the start profile. What irregularity adds are *density fluctuations* in the initial distribution of opinions. They are additional causes for splits and induce opinion changes deep inside the start–profile without any delay right at the beginning of the process. [5]

3 Objections and Criticisms

Central features of the dynamics driven by 1 are understood by the analysis given above. That is the moment—at the latest—in which almost all readers will call into doubt the whole model or certain of its features. In the following I will gather and formulate *some* anticipated objections and criticisms. They all argue that there are *important* facts about real–world opinion dynamics, which are not covered in our basic model.

1. Confidence is often asymmetric: Sometimes it is fashionable to listen more to the right, sometimes everybody listens more to the left.

2. The opinion itself may matter for the asymmetry of confidence: Leftists listen more to the left, rightists more to the right.

3. Fairly often one does know only the opinions of those living close by (in physical or social terms).

4. Simultaneous updating is fairly artificial. Often only two persons or a small subset of persons meet each other and exchange.

5. Whether one takes seriously what someone thinks about one topic often depends on what that person thinks about other topics.

6. Individuals do not get exactly the opinions of others and they make mistakes.

7. The confidence interval of an individual may change over time. Sometimes it may be bigger, sometimes it may be smaller, sometimes with a bias, sometimes without a bias. And even more, the intervals may differ from individual to individual.

8. Applying the arithmetic mean is not the one and only natural way of averaging. Individuals may differ in the type of mean they apply: Geometric,

---

[5]Cf. the examples in [Hegselann and Krause, 2002, Section 4.1].

harmonic, power means of all sorts—to mention only some. The mean they apply may change over time.

9. There are other mechanisms as well: At least sometimes we take seriously the opinions of others independent of the distance of their opinion to ours. Sometimes we want to be different from those around us, especially if there are too many. Intentionally we move away from their opinions.

10. The model misses the most decisive point of rational discourse: there is something like truth, plausibility, and justification that do not play any role in the model.

In what follows I will try to show how one can react on these criticisms *without giving up the idea of radically simplifying modelling*. The strategy will generally be to add simple extensions to our basic model while keeping constant everything else. The extensions *correspond* the objections, starting with objection 1.

4   Extension 1: Opinion Independent Asymmetric Confidence

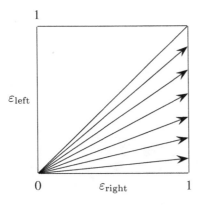

Figure 5. Research strategy for opinion independent asymmetries

Of course, confidence is often *asymmetric*. A *first* type of asymmetry—studied in this section—is *independent* of the opinion an agent holds. Thus, whatever their opinion might be, all agents have the same confidence intervals $\varepsilon_1, \varepsilon_r$ with $\varepsilon_1 \neq \varepsilon_r$. This type of asymmetry can be incorporated as follows: With continuous, one-dimensional, and opinions x taken from the interval $x \in [0, 1]$ confidence intervals $0 \leq \varepsilon_1, \varepsilon_r \leq 1$ make sense. Thus, the parameter space for the confidence interval is no longer the unit interval, but the unit *square*. The diagonal represents the symmetric confidence that we analysed in section 1. Walking along the diagonal was the research strategy underlying the results given in Figure 2. Now again our approach can be a systematic walk through the parameter space. But instead of taking the route along the diagonal we now walk on straight lines *below* (or *above*) the diagonal as indicated by Figure 5. For this type of asymmetry it does not matter whether confidence is biased to the right *or* biased to the left. For all effects we find in that area of the parameter space exist corresponding effects in the triangle above (bias to the left). Therefore it is only *one* of the triangles, either the one below or the one above the diagonal that has to be analysed.

Figure 6 gives the results for *one* stepwise walk below the diagonal, which follows the straight line $\varepsilon_l = 0.25\varepsilon_r$. We make our steps until we get to the point $\varepsilon_l = 0.1, \varepsilon_r = 0.4$. For each value of these 41 steps we repeat the simulation 50 times, always starting with a different uniform distribution. Each run is continued until the dynamics becomes stable. Figure 6 has to be interpreted in the same way as Figure 2.

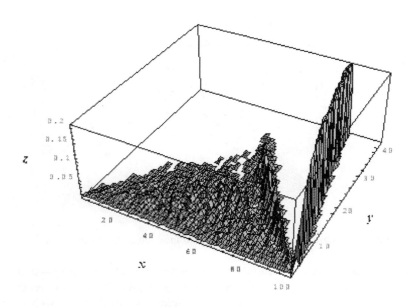

Figure 6. Walking in 41 steps on the line $\varepsilon_l = 0.25\varepsilon_r$

Systematic simulations along the lines described above lead to the following general results:

- As ε_r (and thereby ε_l) increases one always gets into a region of the parameter space where *consensus* prevails. But as ε_l compared to ε_r becomes smaller and smaller the resulting consensus moves into the favoured, i.e. here into the right direction.

- For a very small ε_r and ε_r the dynamics stabilises with a lot of surviving opinions. Thus again we have a phase one might call *plurality*.

- As ε_r and ε_l increases polarisation emerges. As long as ε_l is only a little bit smaller than ε_r it is the type of polarisation we know from the symmetric case:[6] in the left and in the right of the of the opinion space extreme opinion camps emerge, grow, and get closer to each other with an increasing ε_r (and thereby increasing ε_l). It is a somehow a *'symmetric'* polarisation: the camps have about the same size and the same distance from the centre (or the borders, respectively) of the opinion space.

- As ε_r becomes significantly greater than ε_l we observe a new type of *asymmetric* polarisation. The most obvious effect is that a big opinion camp emerges at the right border of the opinion space. This effect is extreme if ε_l is only a small fraction of ε_r (cf. Figure 6). To the left of this main camp—in a certain distance, but still to the right of the centre of the opinion space—we observe smaller but nevertheless outstanding frequencies. This reflects the fact that asymmetric confidence tends to produce in a certain region of the parameter space two or few opinion camps of different size: The bigger one normally more to the border of the opinion space.

5 Extension 2: Opinion Dependent Asymmetric Confidence

The asymmetries of confidence intervals may *dependent* upon the opinion the agents holds. A natural to model an *opinion dependent asymmetry* is to assume that the more left (right) an opinion is, the more the confidence interval of a given *total* (!) size $\bar{\varepsilon}$ is biased direction left (right). For the special case of a 'centre' opinion, i.e. $x = 0.5$, the confidence interval should be symmetric. Thus we introduce an opinion dependent bias β_l to the left and a bias β_r to the right such that $\beta_l, \beta_r \geq 0$ nd $\beta_l + \beta_r = 1$. Afterwards we use β_l and β_r to divide any given confidence interval $\bar{\varepsilon}$ into a left and a right part. Following our intuition stated above, the values β_r should be generated by a monotonically *increasing* function f of the opinion x with $x \in [0, 1]$. By $1 - f(x)$ we get the monotonically *decreasing* function that we need to generate the values β_l—again following the intuition stated

---

[6]Cf. Figure 11a, $\varepsilon_l = 0.9\varepsilon_r$ in [Hegselann and Krause, 2002].

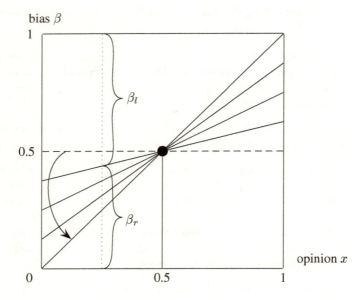

Figure 7. Parameter space for opinion dependent asymmetry

above. Different slopes for the function f would then allow modelling the *strength of the bias*.

Figure 7 illustrates an easy way to characterize this type of asymmetry. The x-axis indicates the opinion. The y-axis or, respectively, the four graphs indicate the opinion dependent *bias* $\beta_r(x)$ to the right and $\beta_l(x)$ to the left. The graphs are generated by rotations around the thick black point $\langle 0.5, 0.5 \rangle$, i.e. according to the function $f(x) = mx + \frac{1-m}{2}$. It is $\beta_r(x) = mx + \frac{1-m}{2}$ and $\beta_l(x) = 1 - \beta_r(x)$. These biases are used to determine how a confidence interval of any given size $\bar{\varepsilon}$ is partitioned into an ε_l and an ε_r. We define $\varepsilon_r(x) = \beta_r(x)\bar{\varepsilon}$ and $\varepsilon_l(x) = \beta_l(x)\bar{\varepsilon}$. Then it always holds that $\varepsilon_l(x) + \varepsilon_r(x) = \bar{\varepsilon}$. It is also guaranteed that $\varepsilon_l(0.5) = \varepsilon_r(0.5)$. In this setting it is the slope m in $f(x) = mx + \frac{1-m}{2}$ that controls the strength of the bias. For $m = 0$ we do not have any bias. Both parts of the confidence interval have, whatever the opinion might be, the same size. As m increases (by rotating the graphs anti-clockwise around $\langle 0.5, 0.5 \rangle$) the bias becomes stronger and stronger: People with a more left (right) opinion listen less and less to the right (left) side of the opinion space. For any positive m it holds that the more one's opinion is located to the left (right), the more one's confidence interval is shifted to the left (right).

This approach offers an easy way to analyse the effects of opinion dependent asymmetries of confidence intervals (Figure 8): for different absolute sizes of confidence intervals we start with symmetry. In each case we let m stepwise increase and study the resulting dynamics by means of simulation. The analysed area of the parameter space is $0 \le m \le 1$.

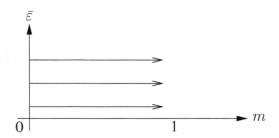

Figure 8. Opinion dependent asymmetries: analysing the parameter space

Figure 9 shows the effects of an increasing bias for $\bar{\varepsilon} = 0.2$. The graphics are of the same *type* as in Figures 2 and 6. The x-axis represents the opinion space $[0, 1]$ divided into 100 intervals. The z-axis represents the average (!) relative frequencies of opinions in the 100 opinion intervals of the opinion space after the dynamics has stabilised. *In contrast* to the former figures the *y-axis does not* represent changes in ε; it now represents *changes of the parameter m* which controls the strength of the opinion dependent bias of $\bar{\varepsilon}$. Along the y-axis m is increasing from 0 to 1 in 26 steps of size 0.04 (while in the former graphics we saw 40 steps of an increasing ε_l or ε_r, respectively). Thus, the graphics represents a walk along *one* of the horizontal lines in Figure 8.

A careful analysis along the lines described above (cf. [Hegselann and Krause, 2002, Section 4.2.2]) supports the following general results:

With an *increasing* opinion dependent bias of confidence intervals

- at least a moderate polarisation *starts earlier*;

- polarisation is getting *stronger* and the polarised opinions are *more to the extremes*;

- it becomes *more and more difficult to reach a consensus*. At a certain point consensus is no longer possible at all. Polarisation is the result instead. If consensus is still feasible, it takes more time to get to consensus.

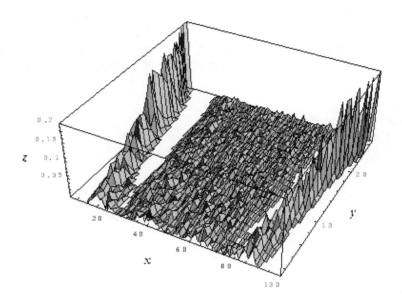

Figure 9. $\bar{\varepsilon} = 0.2$; bias $m = 0 \ldots 1$ in 26 steps

6 Extension 3: Running the Model on Regular or Irregular Grids

Often opinion exchange takes place only among those living close by and belonging to one's *network*, while in our basic model everybody interacts with everybody else. This objection stresses the *local* nature of many interactions which are interactions within a certain *neighbourhood*.

There are several possibilities to cover such a feature. The simplest strategy is to run the model on a low dimensional *grid*. The individuals 'live' in the cells of the grid and interact only with the individuals in the *neighbouring* cells. The grid itself can be constructed in different ways. In the 2-dimensional case one can think of a *regular* construction with a rectangular, hexagonal or triangular structure. Another possibility is the *irregular* structure as given by Figure 10, known as *Voronoi diagram*. To get this irregular structure a certain number of cell generating points (*generators*) is randomly distributed on the plane. The cell around a generator g is the set of all points that are closer to g than to any other generator g'. Following these construction principles results in a structure in which cells—different from regular grids—have different numbers of *adjacent* cells (adjacent in the sense of having a common edge, *not* just a point). The construction principles imply that no cell has less than 3 adjacent cells; at the average cells have 6 adjacent cells. Neighbourhood size can be defined in terms of an upper limit for the number of adjacent cells one has to visit to get to a cell. Figure 10 gives examples for neighbourhood sizes of 2 and 1. The light grey cells are the neighbourhood of the dark grey cells.

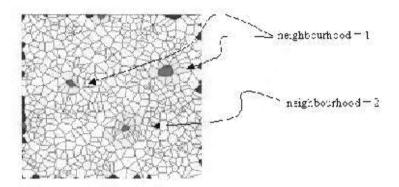

Figure 10. Irregular grid (Voronoi diagram) with different neighbourhoods

To study the effects of *locality* of interaction structures we modify our dynamics defined by 1: we substitute $I(i, x(t))$ by $I_\sharp(i, x(t))$, which is the set of individuals, which i takes seriously *and* live in the neighbourhood of i.

Does locality matter? Figures 11a and 11b give an answer.

Both figures are to be interpreted in the same way as Figure 2 and Figure 6: The x-axis represents the opinion space $[0, 1]$ divided into 100 intervals. The z-axis represents the average (!) relative frequencies of opinions in the 100 opinion intervals of the opinion space after the dynamics has stabilised. The y-axis represents the 41 steps of increasing the confidence interval.

Comparing Figures 2, 11a and 11b gives hints about the effects of locality: The most striking fact is that in 11a, the case in which individuals live in the smallest possible neighbourhood, we do *not* find the Lefty and Righty mountains as in Figure 2 and which indicate polarisation. As neighbourhood size increases—see Figure 11b—these mountains start to reappear. Thus, locality seems to matter a lot. The general *conjecture* is:

- Small neighbourhoods cultivate and foster a plurality of opinions.

- At the same time they seem to prevent the opinion dynamics from ending up with a two-camps-polarisation.

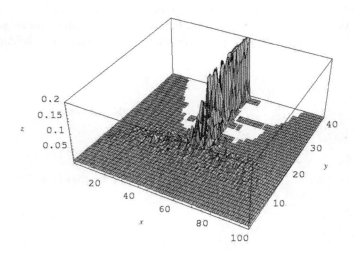

(a) neighbourhood = 1

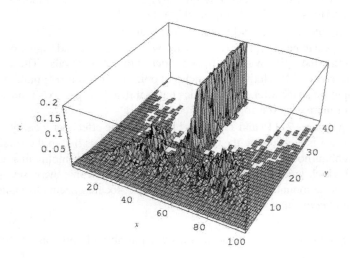

(b) neighbourhood = 2

Figure 11.

A cautious note: This is only a *starting* point for the study of locality since the figures do not say anything about the absolute numbers of surviving opinions, how they are scattered etc.

7   Extension 4: Different updating procedures

Objection 4 calls into question simultaneous updating. Basically updating procedures are idealized interaction structures. From that perspective simultaneous updating according [2] models a type of opinion exchange in which period for period everybody recognizes the actual opinion of everybody else. It may be that sometimes round table discussions, small committees or bodies work like that and, therefore, can be modelled in that way. One may think of the Delphi method as an intentionally designed 'real world' phenomenon where simultaneous updating applies. Reactions on opinion polls published by mass media may be looked at as cases of simultaneous updating. But a major part of opinion exchanges has another structure: Fairly often they are *pair-wise* interactions, for instance, by making a call, speaking to a colleague etc.

The objection makes quite clear that there are alternatives to simultaneous updating. *One* such alternative is pair-wise sequential updating: We select two individuals, which then update their opinions. Supposing an even number n of individuals we select $n/2$ pairs, what implies that simultaneous and pair-wise sequential updating lead to the *same* number of updated opinions *per period*. Figures 12a and 12b indicate that the updating procedure really matters: Both dynamics start with the same start distribution; the homogeneous and symmetric confidence interval is $\varepsilon = 0.1$.

Figure 12 supports *two general conjectures*:

- Compared with simultaneous updating it takes pair-wise sequential updating much more time to converge.

- It seems that for a given confidence interval under pair-wise sequential updating many more opinions survive.

For systematic simulations studies we can regard simultaneous and pair-wise updating as the borderline cases of updating by picking out groups of m out of n individuals for all $2 \leq m \leq n$. The updating groups may be randomly chosen period by period or once forever. Convergence speed and the number of surviving opinions seem to be interesting points of comparison.

8   Extension 5: More opinion dimensions

Objection 5 can be regarded as an attack on the *one*-dimensional opinion space in the basic model. But—and that seems to be true—whether one takes seriously what someone thinks in one dimension is often depending on what that persons

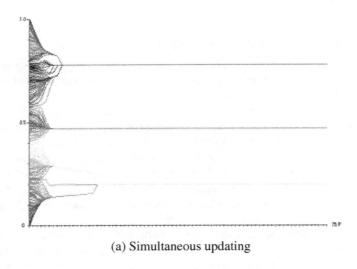

(a) Simultaneous updating

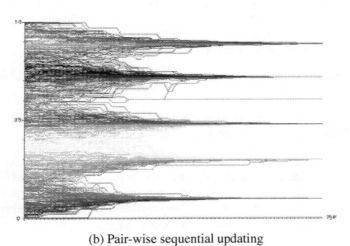

(b) Pair-wise sequential updating

Figure 12. Different updating procedures, same start distribution, $\varepsilon = 0.1$

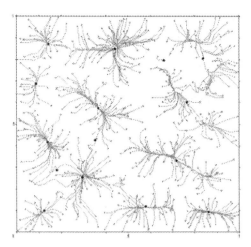

Figure 13. Two-dimensional opinion dynamics. To be taken seriously ε has to be ≤ 0.1 *in both dimensions*. Start with 625 opinions.

thinks in *another* dimension. The natural reaction to that objection is to introduce d opinion dimensions with $d \geq 2$. The persons one takes seriously are those for which their distance to our opinion—now a vector with d components—isn't 'too big' in all dimensions. There are several possibilities to couple the confidence requirement for the d dimensions. A natural starting point could be the requirement that in all dimensions the distance has to be smaller than ε. In the 2-dimensional case that implies to take seriously all those within a certain *confidence square* around one's own opinion, located at the intersection of the diagonals of that square. Requiring that the *Euclidian distance* has to be smaller than ε would mean to take seriously all opinions within a certain *confidence circle* around one's opinion.

Figure 13 gives an example for a dynamics in a 2-dimensional opinion space. Small grey circles are the randomly generated opinions of the uniform start distribution. Dark lines are the trajectories. Thick black circles are the end positions after the dynamics has stabilized.

Simulation runs like that in Figure 13 suggest the *conjecture:*

In higher opinion spaces the emergence of two opposed camps is far less frequent and distinct than in the one–dimensional case.[7]

---

[7] In the meantime Jan Lorenz (Bremen University) has completed his diploma thesis, which analyses in detail the *BC*-model for 2- and 3-dimensional opinion spaces. Cf. [Lorenz, 2003].

9   Extension 6: Random mean

Objections 6–8 attack a whole bunch of assumptions built into the basic model: homogeneity of individuals, symmetry, constancy over time, errorless averaging etc. What the objections underline is heterogeneity, variability, and faultiness.– There are many possible reactions on these objections, for instance introducing some noise with regard to the perception of opinions, introducing a mixed population in which different types of means are applied, fuzzy borders of the intervals, noisy averaging etc. Following this approach one could start to analyse different extensions, each of them involving at least one and often more than one additional parameter. Instead of doing all that, we will make *one* rather *radical and simple* move: We try to cover all the heterogeneity, variability, asymmetry, and faultiness by introducing a *random mean*.

The random mean is defined as follows:

$$(3) \qquad \mu_R = x_{\text{random}} \in [x_{\min}, x_{\max}], \; x_{\min}, x_{\max} \in \{x_1, x_2, \ldots, x_n\}.$$

According to [3] individuals average over a set of opinions by choosing a random value from the interval $[x_{min}, x_{max}]$ of all opinions *within* their confidence interval. This way of averaging is applied by all individuals. The hope is that a homogeneously applied *random* mean can be *interpreted* as modelling a world in which agents are free—period by period—to use one of the infinite number of possible means; and even more, the random mean is supposed to cover faultiness, misconceptions, variable and biased confidence as well.

Figures 14a and 14b show two single runs with the same start distribution and the same confidence interval. The dynamics in Figure 14a is based on the arithmetic mean. It becomes stable in period 9 and ends up with three opinion camps. Figure 14b shows the corresponding dynamics based on random mean. It is obviously approaching consensus though even after 1500 periods the dynamics is still not stable. As it seems under random averaging consensus becomes easier though it takes a long time to get there.

Figure 15 gives the results of systematic simulations organized in the same way as in Figures 2, 6 and 11: The x-axis represents the opinion space $[0, 1]$ divided into 100 intervals. The z-axis represents the average relative frequencies of opinions in the 100 opinion intervals of the opinion space after the dynamics has stabilised. The y-axis represents 41 steps of increasing confidence intervals. For each step we run 50 simulations. But note: Compared with our earlier simulations the step size is now *one order of magnitude smaller*, i.e. 0.001. As a consequence the analysed part of the parameter space ends with $\varepsilon = 0.04$! The simulations are now based on 5000 periods. But even then the dynamics is often not totally stable. At the same time we start getting into trouble with the limits of computational precision of DELPHI, the language used for the simulation program. It becomes

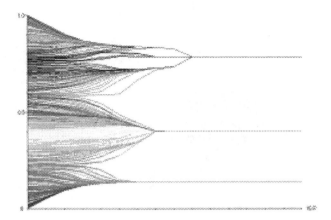

(a) arithmetic mean

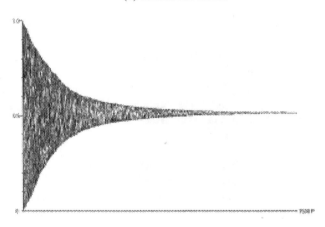

(b) random mean

Figure 14.

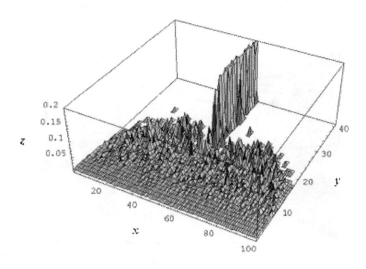

Figure 15. Random mean, $\varepsilon = 0, 0.001, \ldots, 0.04$!!!

undecidable whether the dynamics is either really stable or we are beyond the limits of computational precision.

Figure 16a shows the average numbers of occupied classes after stabilisation as we make our 41 steps (x-axis) of increasing confidence.[8] Figure 16b shows the *coefficient of variation* for the values in Figure 16a.

The simulations support at least three general results:

- The random mean is in terms of convergence speed *much slower* than the arithmetic mean. But for random averaging it takes *much less confidence* to produce effects that the arithmetic produces only with much larger confidence intervals.

- If the random mean captures essential features of *heterogeneity, asymmetry, variability, and faultiness*, then communities of individuals with such features need *much more time, but much less confidence* to get to a consensus.

Heterogeneity, variability, and faultiness seem to *prevent* a community from a dis-

---

[8] Again, while in Figure 15 the classes are 10^{-2} intervals of the opinion space (otherwise one gets problems with visibility), we use for the count in Figure 16a a finer scale, namely 10^{-4} intervals.

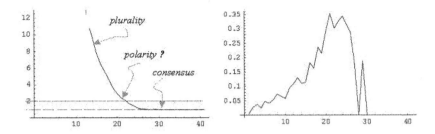

Figure 16. (a) Average number of occupied classes $\varepsilon = 0, 0.001, \ldots, 0.04!!$ (b) Coefficient of variation for figure 16(a)

tinct *polarisation* phase for a certain *interval* of ε-values.[9]

One might question these findings since we apply the random mean in a framework in which the confidence intervals are still symmetric and homogeneous. To clear up the situation we run the random mean simulations under the additional assumption, that the value for ε is only the upper limit for a—period-by-period—*randomly* chosen *left* and *right* confidence interval. Thus, with stepwise increasing confidence intervals 'everything becomes possible'. The striking fact is that these simulations lead to essentially the *same results*. The difference is that the curve in Figure 15a is moved some *few* steps to the right.

10   Extension 7: Other mechanisms, non-conformism, opinion independent competence

Objection 9 stresses the existence of *other* mechanisms that govern opinion dynamics: Different from the bounded confidence model we sometimes take seriously others independent of their distance to our opinion. At least some and (almost all) at least sometimes want to be different from others. Thus they move away from their opinions if there are too many close by in the opinion space. Probably observations like that are part of everybody's experience. Both mechanisms can be integrated in, or better, combined with the *BC*-model.

As to opinion *independent* competence: If one wants to study such a mechanism *without* any interference with other mechanisms—and that is, what one should do *before* combining the mechanisms—then the original Lehrer–Wagner-model is an appropriate framework. The dynamics is given by

$$(4) \qquad x_i(t+1) = \sum_{j \in l} w_{ij} x_j(t), \quad \text{with} \sum_{j \in l} w_{ij} = 1,$$

---

[9]To see that compare Figures 3a and 16a.

where w_{ij} is the weight that individual i assigns to individual j.It is of course another model, but, important to note, within the same framework of *simplistic modelling*.

We can *add* such a mechanism to the bounded confidence model by assigning to each individual i a set K_i of others that i takes seriously independent of whether or not their opinions are within is confidence interval. Averaging on the opinions of all those *either* in K_i *or* within i's confidence interval drives the new dynamics, which can be interpreted as an *interference* of two different mechanisms. To start systematic research we run simulations in a 2-dimensional parameter space, given by, *firstly*, assigning all individuals i a set K_i with an equal and stepwise increasing relative frequency of randomly chosen individuals, and, *secondly*, a stepwise increasing confidence interval.

Redefining and extending the bounded confidence model in a way that it includes non-conformism isn't difficult as well. We could, for instance, define a dynamics in which individual i updates its opinion according to [2] *if* the relative frequency of others within the range $x_i(t) \pm \kappa$ is below a certain threshold σ. Otherwise i makes a random move. To keep it simple we assume homogeneity, constancy, and symmetry for κ, σ, and ε; additionally we suppose correctness of perceptions and calculations. Nevertheless, in a systematic simulation campaign we have to tackle a 3- or 4-dimensional parameter space. The first three dimensions regard κ, σ, and ε. On top of this we have a *fourth* dimension if we want to analyse *mixed* populations in which some update according to [2] while all others follow the redefined mechanism.

11   Extension 8: Epistemic optimism and the attraction of Truth

Objection 10 argues that the model forgets about the most decisive point of rational discourse: truth and justification. Therefore—one could continue the criticism—it can be applied at best to areas where truth and justification do not play any role. One might even hold the view that the model is a formalized version of the much-contested view of social constructivism. Admittedly, truth, justification etc. don't play any role in the model so far. But that can be changed.

There are different theories, conceptions, and understandings of truth. They are based on correspondence, coherence, warranted assertibility etc. Whatever the right theory might be, let's, *firstly*, assume there is a somehow 'objective' truth and, correspondingly, a *true* value T in our opinion space, i.e. $T \in [0, 1]$. *Secondly*, we assume that there is a quest for truth, a *truth-directedness* built into our individuals.[10] But how can we integrate that in our model? We will *not even try* to model explicitly all the processes and actions in the quest for truth (rational ar-

---

[10] See the contribution of Pascal Engel in this volume.

gumentation, reasonable thinking, weighing evidences, making experiments etc.).
What we do instead is to cover all, which might be important, by the *one* simple
assumption that *truth attracts*. Formally that can be done in the following way:

(1) $\quad \bar{x}_i(t+1) = \alpha T + (1-\alpha)x_i(t+1)$, where $0 \leq \alpha \leq 1$.

This dynamics is a convex combination of two components, a first *objective*,
truth-directed component and a second *social* component. In the latter $x_i(t+1)$
is given by [2], our original dynamics. α controls the strength of the attraction of
truth. For $\alpha = 0$ we have our original bounded confidence model in which truth
does not play any role.[11] But for any positive α truth attracts and drives the whole
dynamics in the direction of the true value T. Positive values of α cause in our
model what *epistemic optimism* hopes for, getting closer and closer to the truth.

A *direct* form of truth-directedness and epistemic optimism is modelled by [5].
But one can also think of more *indirect* forms: The start distribution may have a
more or less distinct peak at the truth. The confidence intervals may have a bias
direction truth, the further away the stronger the bias. The size of the confidence
interval may be dependent on the distance to the truth, the closer to the truth the
smaller the interval. Or there may be some noise with a bias direction truth.

Including truth into the model opens a new and more *philosophical field of
research*, driven by questions like: How close and how fast do we get to the truth?

12   The art and heuristics of radically simplifying modelling

There are different types of models, constructed and designed for different purposes. The model developed and analysed above is a model in the tradition of
radically simplifying models. It does *not* aim at *quantitative and exact* predictions
or explanations of real world phenomena. Its purpose is more *a qualitative understanding of mechanisms* in an area where up till now we do not have that much
understanding. Purpose and design of this type of model is totally different from
climate models, models for car crashes or an aero dynamical model for a plane—
all models, which are based on empirical knowledge and well-founded theories.
Our model is a more *explorative* one in an area where almost all is still to be done.
Up till now the empirical bases are some *stylised facts*. One could describe the
whole approach as a *computer-aided speculation* on opinion dynamics.

However, it is—and has to be—a *controlled and systematic* speculation. The
central danger for an approach by radically simplifying models is to drown—
against all good intentions—in *over-complexity*, i.e. to get lost in a high dimensional parameter space. The point of modeling is—at least normally—*not* to illustrate the model's behaviour for arbitrarily chosen values of its parameter space; the

---

[11]The social component could also be given by [4], the Lehrer–Wagner model, or by any other
opinion dynamics that can be built in such a convex combination.

point is rather to get a general understanding of and by the model. That normally presupposes an overall view of what is going on in the parameter space.

To escape from the danger of drowning in over-complexity we implicitly followed (or recommended) in Sections 2 to 11 a certain *heuristics*. Important rules, partly rules of thumb, were the following:

1. Start the analysis with a *minimal model*, i.e. a model that economizes on parameters. Making homogeneity, symmetry, and constancy assumptions is a strategy to keep down the number of parameters. Try to get a perfect understanding of the minimal model. It serves as a benchmark for comparisons with the behaviour of extended and modified versions of the minimal model.

2. Be aware of the *dimensions* of a (minimal or extended) model's parameter space. Wherever there are alternatives to your assumptions, the alternatives constitute a dimension that may matter. The weaker the arguments *either* for fixing a dimension by a specific value *or* for the irrelevance of specific values of that dimension, the more urgent is the analysis of the *whole* dimension. [12]

3. The art of model building is to get a *maximum* of interesting effects out of a *minimum* of assumptions. Thus, when extending a model, the point is *not* to build in whatever seems to be effective 'out there in the world'. The point is to find *smart* extensions, which produce interesting additional phenomena at minimal costs. A smart extension may not even have a directly resembling counterpart in the real world. It may be that it covers in the model the effects of a whole bunch of different real world features (as, for instance, the random mean does).

4. Extensions of a model should be done *stepwise* and with a *manageable step size*. These restrictions are a consequence of the requirement that we have to keep track of what is going on in the whole extension of the parameter space. [13] Point of departure and borderline case of the extension should

---

[12]Our minimal model has—besides the obvious dimensions 'size of the confidence interval', updating procedures, start distributions—two dimensions, which one tends to overlook: (1) The *number of individuals*. In our simulations we get with an increasing confidence the three phases plurality, polarization, consensus. The simulations are run with 625 individuals. We get the same result for about 10 times as much individuals. But it may nevertheless be the case that for *any* confidence interval there is a number of individuals—it may be astronomically high, for example 10^{500}—for which a uniform start distribution normally leads to *consensus*. My personal conjecture is, that that is in deed the case. (2) The opinion space is the *continuous* unit interval. But the opinion space could be *discrete*, what probably matters for the number of surviving opinions.

[13]For beings like us there is an upper limit for the dimensionality of parameter spaces we can overview with ease. With regard to vision, our strongest sense, one has to note that we can easily visualize up to three dimensions. With some tricky and difficult procedures one may visualize two more dimensions. Anyhow, getting an overview becomes increasingly difficult.

be a well-understood state of the model, to allow for an understanding by contrasting.

5. Look whether a complex process can be decomposed and conceptualised as an *interference and overlay of simple mechanisms*. Get first an understanding of the isolated simple mechanisms. Only then start to combine them.

Where did we get following these rules? The *BC*-model (all extensions included) *is not a realistic model*. The model allows for a *detailed understanding of a single mechanism*: revising one's opinion by taking into account the opinions of others that are not too strange—understood in a certain way. Such a mechanism is at work—sometimes and in some. But a mix of mechanisms drives 'real' opinion dynamics. There is, firstly, no guarantee that there are only mechanisms at work, which can be understood by simple models. And there is, secondly, no guarantee that the understanding of interfering, interacting, and overlaying simple mechanisms is within our cognitive reach. But where our single mechanism has the dominant influence and enough data are at hand there we can use the model at least for *qualitative* explanations or predictions.

Acknowledgements

The results presented here are based on joined research and many discussions with Ulrich Krause (Bremen University). This paper draws on Hegselmann and Krause [2002] and Hegselmann and Krause [2003].

BIBLIOGRAPHY

[Abelson, 1964] R. P. Abelson. Mathematical models of the distribution of attitudes under controversy. In N. Frederiksen and H. Gulliksen, eds., *Contributions to Mathematical Psychology*, New York, NY: Holt, Rinehart, and Winston, 1964.

[Chatterjee, 1975] S. Chatterjee. Reaching a consensus: Some limit theorems. *Proceedings of the Internatioal Statistical Institute*, pp. 159–164, 1975.

[Chatterjee and Seneta, 1977] S. Chatterjee and E. Seneta. Toward consensus: some convergence theorems on repeated averaging. *J. Appl. Prob*, **14**, 89–97, 1977.

[Deffuant et al., 2000] G. Deffuant, D. Neau, F. Amblard, and G. Weisbuch. Mixing beliefs among interacting agents. *Advances in Complex Systems*, **3**, 87–98, 2000.

[De Groot, 1974] M. H. De Groot. Reaching a consensus. *J. Amer. Statist. Assoc*, **69**, 118–121, 1974.

[Dittmer, 2000] J. C. Dittmer. Diskrete nichtlineare Modelle der Konsensbildung. *Diploma thesis*, Universität Bremen, 2000.

[Dittmer, 2001] J. C. Dittmer. Consensus formation under bounded confidence. *Nonlinear Analysis*, **47**, 4615–4621, 2001.

[French, 1956] J. R. P. French. A formal theory of social power. *Psychological Review*, **63**, 181–194, 1956.

[Harary, 1959] F. Harary. A criterion for unanimity in French's theory of social power. In D. Cartwright, ed., *Studies in Social Power*. Institute for Social Research, Ann Arbor, 1959.

[Hegselann and Krause, 2002] R. Hegselmann and U. Krause. Opinion Dynamics and Bounded Confidence—Models, Analysis, and Simulations. *Journal of Artificial Societies and Social Simulation (JASSS)*, **5**, 2002. <http://jasss.soc.surrey.ac.uk/5/3/2.html>.

[Hegselmann and Krause, 2003] R. Hegselmann and U. Krause. Collective Dynamics of Interacting Agents when driven by PAM. Submitted.

[Krause, 1997] U. Krause. Soziale Dynamiken mit vielen Interakteuren. Eine Problemskizze. In U. Krause and M. Stöckler, eds., *Modellierung und Simulation von Dynamiken mit vielen interagierenden Akteuren*, pp. 37–51. Universität Bremen, 1997.

[Krause, 2000] U. Krause. A discrete nonlinear and non—autonomous model of consensus formation. In S. Elaydi, G. Ladas, J. Popenda and J. Rakowski, eds., *Communications in Difference Equations*, pp. 227–236. Amsterdam: Gordon and Breach, 2000.

[Lehrer and Wagner, 1981] K. Lehrer and C. G. Wagner. *Rational Consensus in Science and Society.* Dordrecht: D. Reidel Publ. Co, 1981.

[Lorenz, 2003] J. Lorenz. *Mehrdimensionale Meinungsdynamik bei wechselndem Vertrauen.* Diploma thesis, Universität Bremen, 2003. http://www-stuga.informatik.uni-bremen.de/mathematik/archiv/diplome/jlorenz.zip.

[Weisbuch *et al*, 2001] G. Weisbuch, G. Deffuant, F. Amblard and J. P. Nadal. Interacting agents and continuous opinion dynamics, 2001. http://arXiv.org/pdf/cond-mat/0111494.

Rainer Hegselmann

University of Bayreuth

Deptartment of Philosophy

Universitaetsstr. 30

D-95440 Bayreuth

Germany

Email: rainer.hegselmann@uni-bayreuth.de

A Forgotten Common Origin
Comments on Hegselmann

MOSHÉ MACHOVER

I must begin with an apology: I am delivering these comments under almost false pretences.

Let me explain. Donald Gillies' kind invitation, asking me to undertake this task, was issued in December 2002, when this conference was at the planning stage, before the papers presented here were available. At that stage, Rainer Hegselmann apparently intended to include in his paper a section about the application of his model to evaluating various voting procedures. According to an explanation I received from Donald, in this projected section Rainer Hegselmann would regard the group of agents as a committee that has to come to a decision about a certain value. To this end, the committee could apply various voting procedures. It would be easy to calculate period for period the outcome for a major set of procedures. By this type of analysis one would be able to gain fresh insights about voting procedures, their advantages and disadvantages, depending for instance on the time the committee has at its disposal. Donald Gillies recruited me to comment on Rainer Hegselmann's paper because for quite a few years I have been working in the field of social choice, the mathematical study of collective decision making.

But, as often happens in such cases, in writing his paper Rainer Hegselmann discovered that it was getting longer than originally planned. The plan had to be trimmed; and the section about application to voting had to be omitted.

So I found myself committed to commenting on a paper on whose subject-matter I cannot claim to have any special expertise.

Fortunately my embarrassment is not complete: I discovered that between the topic addressed by Rainer Hegselmann and my own specialist field there is nevertheless some unexpected connection—thanks to which I have allowed myself to insert the modification 'almost' in my opening sentence.

The unexpected connection is a forgotten historical one, which I think deserves to be better known.

Rainer Hegselmann's approach can perhaps be regarded as a quasi-epidemiological modelling of the evolution of opinion within a population. He mentions some earlier antecedents of his model; but there is an earlier one—

Laws and Models in Science, 47–49.

perhaps the earliest—that is absent from his list. This is the work of Lionel Penrose.[1] Unfortunately, some seminal scientific contributions of this extraordinary mathematician, geneticist and psychiatrist were subsequently forgotten and were later reinvented by others, to whom they are now generally credited.

As it happens, Lionel Penrose was also a pioneer in my particular topic of research: the measurement of voting power. In fact, his [1946] paper is—as far as I know—the very first scientific treatment of this topic. His work on this topic was evidently motivated by the establishment of the UN and the hopes—widespread at the time—that it would evolve into a 'world government'. The dashing of these hopes with the onset of the Cold War may partly explain why his work was not taken up by others at the time and was forgotten. It took about 20 years for some of his ideas to be reinvented independently by J. F. Banzhaf, and even later by J. S. Coleman and others. The measure of voting power originally proposed by Penrose is generally known as the *Banzhaf* (or *Coleman–Banzhaf*) *index*.[2]

Penrose returned to this topic a few years after that paper, in [1952]. In this extraordinary little book—a mere 73 pages—he ranges over a great variety of topics, among which is that of voting power. But most of the ideas presented in this booklet concern the epidemiology of the spread of ideas.

In order to convey the remarkable range of topics covered in Penrose's booklet, I reproduce below its Table of Contents.

I mentioned that *some* of Penrose's ideas on voting power were later reinvented by others. But over the years of my own work in this area I have found that there are still some further gems that have lain hidden in his writings.

I suspect that the same may be true of his contributions to the epidemiology of ideas. I therefore end with a plea: have a good look at his writings on this topic; you may well find there valuable insights that have not been sufficiently utilized as yet.

---

[1]Lionel Sharples Penrose (1898–1972) studied mathematics at St John's College, Cambridge; awarded degree of Moral Sciences Tripos (Cambridge, 1921); did a medical course at St Thomas's Hospital, London, and was awarded the degree of MD, (1930). His main research was in genetics as well as psychiatry. Together with his son, the famous mathematician Roger Penrose, he invented the Penrose triangle and the Penrose stairs, impossible objects popularized by the artist M. C. Escher.

[2]Dan Felsenthal and I discuss the convoluted history of the study of voting power in our book [Felsenthal and Machover, 1998], and more fully in a forthcoming paper [Felsenthal and Machover, 2005].

CHAPTER		PAGE
I.	Introduction	1
II.	Effects of size of the group	6
	(a) Formal consequences	
	(b) Stratified control	
	(c) The power of one vote	
	(d) Conclusion	
III.	Analogy between mental and physical epidemiology	13
IV.	Analysis of crazes	18
V.	Outbreaks of religious enthusiasm	23
VI.	Panic reactions	28
VII.	Politics	36
	(a) General principles	
	(b) Two-party system	
	(c) Equitable representation	
VII.	War	47
IX	The transmission of ideas	55
	(a) General principles	
	(b) Infectious properties of ideas	
	(c) Pathological ideas	
	(d) The condition of the recipient	
X.	Concluding remarks	66
	Bibliography	69
	Appendix	71

BIBLIOGRAPHY

[Felsenthal and Machover, 1998] D. S. Felsenthal and M. Machover. *The Measurement of Voting Power: Theory and Practice, Problems and Paradoxes*. Cheltenham: Edward Elgar, 1998.

[Felsenthal and Machover, 2005] D. S. Felsenthal and M. Machover. Voting power measurement: A story of misreinvention. Forthcoming in Social Choice and Welfare, 2005.

[Penrose, 1946] L. S. Penrose. The elementary statistics of majority voting. *Journal of the Royal Statistical Society*, **109**, 53–57, 1946.

[Penrose, 1952] L. S. Penrose. *On the Objective Study of Crowd Behaviour* London: H. K. Lewis and Co, 1952.

Moshé Machover

Department of Philosophy
King's College London, Strand
London WC2R 2LS, UK

Email: moshe.machover@kcl.ac.uk

Models and Modals

HUW PRICE

1 Introduction

Pragmatists recommend that in approaching a problematic concept, philosophers should begin by examining the role of the concept concerned in the practical, cognitive and linguistic life of the creatures who use it. I'm interested in pragmatic accounts, in this sense, of the various modal notions we encounter in science—causation, probability, counterfactual conditionals, and so on. In this paper, I want to propose that these accounts should avail themselves of the vocabulary of theoretical models.

Although my concern is thus with the application of models to the study of modals in general, I have a special interest in the case of causation. In previous work, I've defended an 'agency' or 'manipulability' approach to causation. This approach links our possession of causal concepts to the fact that we are agents. In the version that I prefer, it is a pragmatic account, in the above sense. (It is also a perspectival account, in a sense I'll be trying to clarify further below.) Some writers (e.g., [Pearl, 2000; Woodward, 2001]) agree about the centrality of notions of agency and manipulation to an understanding of causation, but take the resulting view in a more realist or objectivist spirit. From my point of view, then, there are two groups of opponents—those who need convincing about the centrality of agency and manipulation in an account of causation in the first place, and those who need convincing only about the pragmatic or perspectival character of the best such account. In both cases, however, a pressing task is to clarify the perspectival option. That's what I'll be attempting here, with the aid of theoretical models.

Clarifying the perspectival option is a matter of locating it on philosophical maps—or, in some cases, redrawing the maps so as not to exclude it by default. Philosophers have a tendency to think of the question 'Is there any such thing as causation?' as on all fours with 'Are there any Tasmanian tigers?' or 'Are there magnetic monopoles?' This has the effect of simply excluding some views of causation, views according to which there are important differences between the question about causation and the questions raised by physics and natural history.

As an example of someone who appears to be missing the relevant possibility, consider this famous remark from Bertrand Russell:

Laws and Models in Science, 51–71.

> All philosophers, of every school, imagine that causation is one of the
> fundamental axioms or postulates of science, yet, oddly enough, in
> advanced sciences such as gravitational astronomy, the word 'cause'
> never occurs ... The law of causality, I believe, like much that passes
> muster among philosophers, is a relic of a bygone age, surviving, like
> the monarchy, only because it is erroneously supposed to do no harm.
> [Russell, 1913]

Russell seems to be arguing that physics has shown us that there is no such thing as causation—a discovery about what the world contains, apparently, on a par with those of natural history and physics itself. But I want to show that there's an important option that Russell thus overlooks. Extending Russell's own metaphor, I'll call it *causal republicanism*.

Consider the political case. When we reject the view that political authority is vested in our rulers by god, we have two choices. We can reject the notion of political authority altogether, or we can regard it, as republicans do, as vested in our rulers by us. The republican option exists in metaphysics, too, where it is an alternative to realism and eliminativism. In the case of causation, it is the view that although notions of causal power are useful, perhaps indispensable, in our dealings with the world, they are a category constructed by us, not provided by God.

In comparing causation dismissively to the monarchy, Russell seems largely blind to this republican possibility—a failing he shares, I think, with many of his realist opponents. In my view, thinking of eliminativism as the sole alternative to causal realism is like thinking of anarchy as the sole alternative to the divine right of kings. Thus I agree with Russell in rejecting a certain kind of realism about causation, but disagree about the relevance to this conclusion of the issue of the eliminability of causal notions from physics. For a republican, causation may turn out to be both ineliminable and anthropocentric. In my view, the best versions of the agency approach give causation this republican flavour.

As I noted above, my interest in agency accounts of causation is part of a broader concern with pragmatic approaches to the various modalities employed in science. There are deep connections among the modal notions, I think, and hence much to be learnt by considering them as a group. More importantly for present purposes, the points I want to make are more easily made for probability than for causation, the relevant landscape being simpler and better-known. In particular, there's a familiar debate about the relevance of physics to the question as to whether there are (non-trivial) chances or objective probabilities—in some ways, a probabilistic analogue of the issue raised by Russell. I think that that debate, too, often misses important parts of the landscape. But it's easier to explain why than for causation directly, because the landscape is so familiar. Dialectically, then, it makes sense to begin with probability.

Most of the paper will thus be concerned with probability. In the next section, I'll distinguish three conceptions of what we are doing in *modelling* probabilities. There are two aspects of this approach to flag at this stage, the first the three-way distinction itself, and the second the fact that it is couched in terms of the functions of theoretical models. The relevance of the three-way distinction is that one of these three conceptions is easily overlooked, and yet crucial, in my view, both in deciding what's right and what's wrong about Russell's claim, and in understanding the nature of modal perspectivalism.

As for the focus on models, I've said that I'm trying to call attention to, and clarify, a kind of pragmatic perspectivalism about modal notions. It would be possible to do this by talking about the functions of theoretical language. Expressed this way, however, the concerns of pragmatism and realism are apt to seem orthogonal. Pragmatists focus on language, realists on reality, and the two sides can easily seem to be talking past one another. Focussing on models makes it easier to find common ground. As we'll see, the kinds of things a modal pragmatist wants to say find natural expression as theses about the functions of models (for creatures in our situation). While on the realist side, issues about the use and role of theoretical models are already sufficiently in play to present the questions the pragmatist wants to raise as 'more of the same'. The result, hopefully, is a more accessible and more 'scientific' modal pragmatism—and a pragmatism with a less linguistic face, via models with a more human face. [1]

2   Probability—Models and Meta-models

Let's begin, then, with a question about probabilistic models. What are we modelling when we model probabilities? Or, more neutrally, what is the function of probabilistic models—what do we use them *for*?

I want to distinguish three different conceptions of the role of probabilistic models, especially in science. In each case, I want to be able to think about the relation of the models in question to the needs and characteristics of the *users* of those models. I want to be able to consider the relevance of variations in characteristics *of the users,* to the utility and possibility of their use of probabilistic models, under particular conceptions of what those models involve. In order to think about these issues in the abstract, idealised way typical in science, I want to be able to model them. So I'll need what I'll call *meta-models*—models of the users of probabilistic models. Here, then, are three possible views of the functions of probabilistic

---

[1] One important aspect of this shift from language to models is that it helps to distance us from naïvely representational conceptions of the role of theoretical language in science. Pragmatists are often foes of representationalism. While theoretical models are not inevitably conceived in anti-representational terms, it is uncontroversial that they can have non-representational functions (see, e. g., [Morgan and Morrison, 1999]). Language wears on its face a representational complexion. Pragmatists want to argue that this complexion is no more than skin-deep—not a reliable guide to underlying function and structure—but the case is much easier to make if we start further from the surface.

models, together with some remarks about the associated meta-models.

2.1   Objectivist models

On this view our aim in modelling probabilities is to model an aspect of the (modeller-independent) physical world. Probability is regarded as an aspect of the objective world, on a par with other features studied by physics. There are different views about what such objective probabilities are, of course. The options include propensities, hypothetical limiting frequencies of various kinds, and theoretical entities, not further specified, but picked out in virtue of their relevance to our decision-theoretic psychology—that aspect of the world, knowledge of which makes rational certain degrees of belief.

If probabilistic models are understood in this objectivist way, what can we say about the users of such models? How should we model such creatures, in our meta-models? There are two crucial points. First, the modelled creatures need to be modelled as *representers*—creatures whose modelling aims to 'mirror' some aspect of the world they inhabit. Second, their use of models of probability has to be seen as *attempting to represent* in this way—the models themselves must have a representational function. (The second point serves to rule out the case of users who, while representers in other ways, use their probabilistic models for some non-representational purpose.)

2.2   Subjectivist models

An alternative view is that in modelling probabilities our aim is to model psychological states—credences—to some feature of which the probability calculus is applicable, at least under idealisation. At least in a loose sense, then, this view holds that when we model probabilities, we model something subjective—some feature of our own minds.

What meta-models are appropriate in this case? As in the objectivist case, the modelled creatures need to be modelled as representers, whose modelling aims to mirror some aspect of the world they inhabit. In this case, however, the feature represented is part of their own psychology. Inter alia, then, they need to be creatures with the relevant psychology—creatures with credences. Note that this wasn't necessary in the objectivist case. If probabilities are part of mind-independent reality, then in principle creatures without credences ought to be able to model them, even if the part of reality thereby modelled needs to be characterised in terms of its relevance for creatures who do have credences.

A detailed subjectivist meta-model may be expected to tell us, among other things, how the relevant psychological states vary with perceived features of the creatures' external environment. (We're assuming here that the function of probabilistic models isn't just navel-gazing—somehow, that self-descriptive psychological modelling has some wider point.) Presumably such things as observed relative frequencies will be relevant at this point. But if subjectivism is to retain its main advantage over objectivism—that of avoiding the metaphysical and epistemological 'queerness' of modal facts, in favour of something commonplace, though psychological—these features had better be non-modal.

One point to emphasise. In the sense in which I'm using the term, subjectivist models of probability are 'self-descriptive'. They are models of the credences of an agent—of the user of the model, in fact, or some idealisation of the user. It is important to note that not all so-called subjectivist accounts of probability need be subjectivist *in this sense*. Some may be closer to the perspectival view I'm about to describe. And some, perhaps, may simply fail to distinguish between the two options.

2.3 Perspectival models

The third possibility is that what we model when we model probabilities is neither some mind-independent aspect of reality, nor an aspect of our own psychology, but rather the world 'as it looks' from the standpoint of such a credence-based psychology—we model a 'projection' from such a psychology. Of course, more needs to be said to make this notion precise. I've just used two metaphors, one visual and one projective. These are neither obviously compatible, nor the only ones in play in this area. Later I'll introduce a third metaphor and say something about each, and their connections.

For the moment, however, we have enough to note some important points about the meta-model associated with this perspectival conception of first-order models. On the one hand, it has something significant in common with the subjectivist case, in that the users modelled in our meta-model need to be endowed with credences. On the other hand, there is something important that differs compared to both the previous cases, in that the users are not modelled as *representers*. The function of their first-order models is not representational.

One question we might appeal to meta-models to address is that of the utility of such non-representational models for the users concerned. To address such a question, we need to include enough 'environment' in the model, and enough detail concerning the users themselves, to explain how such modelling contributes to the well-being of such a creature in such an environment. [2]In principle this question can be asked in the representational cases, too, of course, though there we might expect it to have been addressed at a higher level of generality, rather than specifically with respect to probabilities.[3]

Thus we have three conceptions of the nature of probabilistic models, and three

---

[2]Strictly, the perspective admits a range of options at this point. One, for example, is that the modelling has no particular advantage.

[3]Though it is relevant here that some standard answers seem inapplicable in some modal cases, given certain views of the nature of modal facts. There may be a puzzle about why modal beliefs are useful, for example, if they represent causally isolated possible worlds.

associated meta-models of the relation between such models and their users. This
allows us to ask the question, which meta-model best fits *us?* Notice that this is
an empirical question—a roughly formulated empirical question, to be sure, but in
principle a matter to be investigated by science (indeed, by the human sciences,
for it concerns an aspect of human linguistic behaviour and psychology).

Suppose it turns out that the perspectival option offers the best answer to this
question. Then it would seem to be a mistake to regard 'Are there really probabilities?' as itself an empirical question, on a par with 'Are there Tasmanian
tigers?' or 'Are there magnetic monopoles?' Why? Because to read the question
about probabilities in this way is to *presuppose* a representationalist conception of
the functions of probabilistic modelling—the conception rejected in perspectival
meta-models. In other words, it is to presuppose that what we are trying to do with
our models of probability is the same kind of thing, roughly speaking, as we are
trying to do with our models of Tasmanian megafauna or the quantum world, viz.,
to represent aspects of our external environment.[4]

In order to address the question as to which meta-model best fits our own practice, however, we need a better sense of what the perspectival option involves. Our
next task is therefore to clarify the notion of a perspectival model.

3   What is a Perspectival Model?

Let's begin with three metaphors for perspectivalism, a visual metaphor, a projectivist metaphor, and a fictionalist metaphor. By combining elements of these three
metaphors, I want to bring into focus what I take to be the core of an interesting
perspectivalism.

3.1   The visual metaphor

The visual metaphor thinks of a perspective as like an aspect of reality *viewed from
a particular standpoint.* Here standpoint usually means a spatial standpoint—what
is perspectival is the view from a particular spatial location and orientation. But
in principle it might include aspects of the observer's 'location' in a more general sense, such as aspects of her visual system. (The view through rose-coloured
spectacles might thus be thought of as a particular perspective, for example.) This
metaphor makes it is easy to see how models can be perspectival. We just think of
a model as containing only what is visible from the viewpoint in question. Moreover, because the viewpoint in question is straightforwardly observer-dependent,
the metaphor gives us an easy model of dependence on a contingent feature of an
agent's circumstances—their spatial location, the colour of their spectacles, and so

---

[4]I think this issue is actually more subtle than I here make it sound, because the contrast between
representational and non-representational uses of models less clear-cut than this formulation assumes.
However, I think the contrast between the functions of probabilistic modelling and other kinds of scientific modelling survives a more careful formulation.

on.

On the other hand, the visual metaphor suggests that a perspective is merely an observer-dependent selection from a set of things which are—in themselves, so to speak—observer-independent. This wouldn't be true of all versions of this kind of perspectivalism—it is explicitly not true of Russell's [1914] construction of objects from perspectival sensibilia, for example. But where it is true it is unhelpful. The more interesting cases are ineliminably perspectival, in the sense that we can't achieve a non-perspectival description simply by including more.

3.2   The projectionist metaphor

The classic statement of the projectionist metaphor comes from Hume's distinction between the operations of reason and taste:

> Thus the distinct boundaries and offices of *reason* and of *taste* are easily ascertained. The former conveys the knowledge of truth and falsehood: The latter gives the sentiment of beauty and deformity, vice and virtue. The one discovers objects as they really stand in nature, without addition and diminution: The other has a productive faculty, and gilding and staining all natural objects with the colours borrowed from internal sentiment, raises in a manner a new creation. [Hume, 1998, p. 163]

This metaphor has the great advantage, in my view, of calling attention to distinctive aspects of our psychology, relevant to the perspective in question—*that from which we project,* in effect. In the probability case, we know what this feature is: credence, in its raw or idealised form. But unlike subjectivism proper, which regards probabilistic models as models *of* our credences, this approach regards them as models of the projections of these psychological states—of the 'objectifications' of the credences, the 'new creations' with which our faculties gild reality.

Of course, this remains rather metaphorical, and what it means needs clarification. It might be phenomenological, for example, as it seems to be in Hume. It might be cashed out in more linguistic terms, as the construction of a practice of making claims and reasoning in ways which ultimately 'express' these credences. Or it might be cashed in the 'model model' itself, so that what projection really amounts to is the construction of models whose ontologies stand in the appropriate relationship to the inner states concerned. More on these options later. (It seems to me likely that the latter two options go hand-in-hand.)

For now, I want to call attention to what seem to me the twin advantages of the projectionist metaphor. Firstly, in calling attention to a distinctive aspect of our psychology from which we 'project', it identifies a contingent feature of the speaker that *grounds* the perspective in question. The identification of such a feature is absolutely central, in my view, to any interesting perspectivalism of this kind. Indeed, it is what makes it perspectival, for it is variation with respect to this 'contingent ground' that constitutes variation of perspective.

Note that we shouldn't assume that the relevant ground will always be psychological. In other cases it might be spatial or temporal location, for example. Or it might be location on some non-spatiotemporal scale of variability. (Some crude examples: when we call things hot or cold, tall or short, we are comparing their temperature or height to our own.) Whatever it is, the first important virtue of the projectionist metaphor is that it emphasises that perspectival modelling is a game creatures are equipped to play, in virtue of the fact that they possess or occupy the contingent ground in question.

The second virtue of the metaphor is its comparative clarity on the important issue of what distinguishes perspectivalism from subjectivism. Projection is coloured by 'internal sentiment' (or its analogues in other cases), but it is not a representation of those internal sentiments.[5]

Taken together, these two characteristics comprise the core of the most interesting notion of perspective, in my view. On the one hand, perspectival models (or concepts, or judgements) depend on some contingent feature or ground, possession of which is a precondition of use of the model, concept or judgement in question. On the other hand, such models, concepts or judgements don't *represent* that feature or ground, explicitly or implicitly. (The contingent ground of a perspectival judgement is always *backgrounded,* as we might say.)

3.3   The fictionalist metaphor

This metaphor has the advantage of helping to emphasise that the perspectivity belongs to the background, not the foreground. From within the perspective in question, its objects don't look perspectival. They simply look like objects. This is obviously the case in fiction—it is (almost always) inappropriate, within a fictional context, to portray its objects as fictions. Our fictions don't say, 'It was a dark and stormy and fictional night', except occasionally for self-consciously convention-busting effect. The label 'fictional' is imposed from the outside, when we comment on the status of such objects.[6]

---

[5]However, more needs to be said at this point about the nature of the distinction—especially by someone who, like me, is inclined (see, e.g., [Price, 2004]) to deny that there are any genuine representations, in anything more than a deflationary sense. The distinction requires careful attention to the relations between models and judgement, and to the differing assertion and rejection conditions for the judgements associated with self-descriptive and projective models.

[6]Note that we have a choice about how we put the observer into the fictionalist meta-model. We can put the observer in the fiction, modelling her as a representer of the objects which are fictional from our point of view but not from hers. Or we can model her as a user of (rather than a participant in) the fiction in question. If the meta-model is to represent ourselves, then clearly the latter approach is the right one—we don't want to model ourselves as merely fictional. But the former approach is also useful for some purposes. In particular, it gives us a way of thinking about what the perspective is like 'from the inside'—from the standpoint from which its fictional character is not apparent. How it seems

Of course, we don't want to banish perspectivity for ever. We want to be able to see it *as theorists,* for otherwise perspectivalism would have nothing to say. It gets back in when we ask why a particular fiction should be useful for *these* creatures in *those* circumstances. Perhaps a disadvantage of the fictionalist metaphor, compared to the projectionist metaphor, is that it does nothing to direct our attention to this issue. It would be easy to think of all fictions as on a par, and to fail to notice the ways in which particular fictions may be adapted to—indeed, dependent on—particular needs and circumstances.

A better strategy is therefore to combine this third metaphor with the second—to think of projection as production of fictional ontology, riding on the back of the various psychological and other commonalities that constitute the contingent grounds of the perspectival models in question. We thus ask, why is it useful for creatures occupying those grounds to invent such models? (Why ontology? And why talk of truth?) To make this work, fictionalism needs to have the flexibility to connect particular aspects of the fictions concerned to the relevant aspects of the user's circumstances. Uniquely among all fictions, for example, the fiction of chances needs to connect with credences in an appropriate way—a way that looks (from the inside, as it were) like Lewis's Principal Principle [Lewis, 1980].

A further strategic advantage of the fictionalist metaphor is the way in which it connects with familiar views about the functions of theoretical models in science. It is already well-recognised that models that are in some sense fictions may nevertheless play an important role in science. One version of this view is instrumentalism, which rejects a representationalist conception of the role of theoretical models altogether. Another version, less radical, recognises an important role for models embodying fictional idealisations in the context of a generally realist view of scientific theories. In a sense, my modal perspectivalist simply wants to give these ideas an extra degree of freedom—to suggest, for example, that the utility of instrumentalist models is a more complex matter than usually assumed, and may rely on particular contingent features of the users of those models, such as the fact that their psychology includes credences. My view thus compares to a kind of multifunctional fictionalism. Models are tools, and the kinds of tools we need depends on the kind of creatures we are.

The fictionalist metaphor has one significant disadvantage, from my point of view. Roughly, it is more anti-realist than it needs to be (or than I want to be). I'll return to this point at the end of the paper. Until then, I'm happy to ride in tandem with fictionalist views.

Summing up, then, we have three guiding metaphors for perspectivalism: the metaphor of visual perception from a particular viewpoint, the projectivist metaphor,

---

to us is just how it would seem to those fictional creatures, who perceive real chances. In this case both the chances and the perceivers of chances are fictions, but fictions which tell us a lot about our own real phenomenology.

and the fictionalist metaphor. I've suggested that the second and the third are more useful than the first, and all the more so if they are combined, so that the source of the projection explains the genealogy and utility of the relevant fictions.

4   Perspectival Models in Science?

A republican view of causation or probability would agree with Russell that there is a sense in which these things are not among the constituents of the world discovered by physics, yet disagree that they should be banished from science. Republicans contend that despite their human origins, these modal notions may play a deep role in science. But how could this be? Isn't science supposed to reveal the world as it appears 'from nowhere', rather than the world as it appears from some particular human perspective?

Two initial responses to this concern. First, I want to note, but set aside for time being, the possibility of arguments that there can be no such thing as a non-perspectival description, a viewpoint genuinely 'from nowhere'. I set this aside not because I believe that no such argument is available. (On the contrary, as I'll explain.) But our present interest is in the contrast between modal and non-modal perspectives, and a global argument would be blind to this contrast.

Second, even if there were a view from nowhere, and it were the job of science to describe it, it might nevertheless be helpful to distinguish between 'pure' science, which did just this, and 'applied' science, which was allowed to be perspectival in various ways, in the service of distinctively human interests. Given such a distinction, it would be a legitimate question whether chance, causation, and the like fall on one side of it or the other. To understand the question, we need to be able to bring into focus the perspectival option. In other words, we need to be at home with the idea of modelling reality as it appears from some distinctively human perspective.

Again, the probability case is helpful. On the one hand, it is a case in which the philosophical landscape is sufficiently well-mapped for it to be relatively uncontroversial that non-objective probabilities have some place in science. On the other hand, it has enough connections to modality in general to serve as a gentle introduction to the possibility of a broader perspectivalism.

5   Perspectival Probabilities

The analogue of Russell's claim for the case of objective probability would be that physics has shown that that are no such things. It is widely believed that this claim is false, because quantum mechanics has shown that there are objective probabilities. In the background here, however, is the view that this analogue of Russell's claim would have been true, if physics had turned out to be deterministic; and would still be true, if the right version of quantum theory (or its successor) turned out to be deterministic (as in Bohm's theory, for example, and other no-collapse interpretations of quantum mechanics). Even now, then, we should concede that Russell might turn out be right about chances—or so goes the orthodoxy.

However, it's not hard to see that the orthodoxy can't be the last word on whether probability has a serious role to play in science. It would be a scandal if the issue of the legitimacy of uses of probability and probabilistic models elsewhere in science (and other aspects of everyday life) couldn't be settled until we knew whether the final microphysics was deterministic. So all these other uses of probability need to be explained in some other way.

Let's put this in terms of meta-models. Recall our three hypotheses from Section 2 about how we should model our own use of probability. The objectivist meta-model models us as creatures detecting and representing objective chances— features of the mind-independent world. The subjectivist meta-model models us as creatures representing features of our own psychology. And the perspectival meta-model models us a creatures modelling probabilities in some other non-representational mode (perhaps fictionally, for example). We know that the first hypothesis is a bad hypothesis, at least for all the uses of probability which don't show the right kind of sensitivity to the question as to whether final physics is deterministic. That leaves subjectivism or perspectivalism. The choice between subjectivism and perspectivalism isn't clear-cut, perhaps, and can't be, until perspectivalism is well-understood. So for time being let me lump them together, and use the term republicanism for both.

Thus the choices that Russell seems to allow in the case of causation— eliminativism or objectivism—can't be the only options for science in the case of probability. In that sense, probabilistic republicanism is not only on the table but dominates the table, in a way that it takes a perverse fixity of gaze to fail to see. And once it is seen, it ought to be seen as a threat. Whatever the right republican story in all the cases insensitive to final physics, what is to prevent it working in that case, too? After all, whatever it appeals to—our epistemic situation, for example—is also going to characteristic of the latter cases, apparently (if anything more so, because of the limits physics places on our knowledge). So the republicans at the table seem in danger of making the objectivists redundant.

This redundancy argument could be couched in terms of meta-models. The relevant comparison would be between (i) a meta-model of idealised epistemic agents detecting probabilistic states of affairs, which they then connect to relevant features of their own psychology; and (ii) a meta-model of creatures beginning with such a psychology, and such an epistemic situation, and then finding it useful to construct perspectival models in the appropriate relation to their psychology. It seems that even an objectivist about probability has to admit that the second meta-model is appropriate to all those cases, including cases in science, in which (by the objectivist's lights) there are no chances. It is hard to see how an appeal to a combination of simplicity and Occam could fail to recommend that the second model be applied universally. In other words, it is doubtful whether science *ever* needs genuinely objectivist models of probability.[7]

Perspectivalism has other attractions in this case. Beginning on the psychological or pragmatic side, it makes easy work of things that are often difficult for objectivists, viz., accounting for the epistemology and relevance of beliefs about probability. If there are chances, how can we know about them, and why does that knowledge have the significance it does for decision?[8]

It might be objected that there's a huge countervailing disadvantage. As the objectivist might put it:

> What about the probabilities in quantum mechanics? Surely you're not going to claim that those are perspectival? What could be less perspectival than quantum mechanics—which, after all, is surely the best candidate we have for an account of what the world looks like 'from God's perspective', or 'from nowhere'?

For my part, I'm happy to bite this bullet. I'm happy to accept that at least as standardly formulated, quantum mechanics embodies a perspectival description, a description tailored to (an idealised version of) our own perspective—that of creatures needing to predict and act for an uncertain future, on the basis of knowledge of the past. In particular, I think that this is the right way to think of the probabilities in quantum mechanics.

'But then what is reality *really* like?', my opponent might ask. At this point, there are a number of options. One would be to argue that the question is in any case impossible to answer, for familiar reasons. It's a familiar idea, accepted by many realists, [9] that we know the world only under some structural description. But what goes into a structural description? If structure is always *inter alia* modal structure[10]—if we know reality only in its causal and dispositional aspects—and

---

[7]For an interesting recent argument to a similar conclusion, see [Strevens, 2003, Section 5.6]. I am grateful to Adam Elga for calling this to my attention.

[8]These are sometimes called the problems of upward and downward inference. They are closely related to what van Fraassen [1989] calls the identification and application problems.

[9]See [Langton, 1998] and [Lewis, forthcoming]—and also [Worrall, 1989; Zahar, 1996; Ladyman, 1998; French, 1998; French and Ladyman, 2003], for similar views in the philosophy of science. Although all these authors agree, roughly speaking, that we know the world only in its structural aspects, French and Ladyman differ from the others in thinking, in effect, that structure is all there—that fundamental ontology lies at the structural level. In their version, then, there is no world from which knowledge of structure alone leaves us cut off, but nor, therefore, is there any room for the present objection to my modal perspectivalism. From my point of view, then, the two versions of this structuralist view are equally useful.

[10]Cf. this recent characterisation of structural realism by French and Ladyman:

> [T]here is a minimal metaphysical commitment that we think structural realism ought to entail. This is that there are mind independent *modal* relations between phenomena (both possible and actual), but these relations are not supervenient on the properties of

modal properties are perspectival in the way I'm suggesting, then accepting that quantum mechanics is perspectival in this way is not a reason for thinking of it as second-rate, compared to some achievable but less perspectival theory.

The realist opponents I have in mind don't think that causal and dispositional properties are perspectival, of course. My point is that they can hardly object that a perspectival view of the modal properties on the grounds that it leaves us cut off from the world in itself, because they too think we are thus cut off. So if I had to live with the conclusion that we could know nothing not 'tainted' with modal perspectivity, I could do so. And I'd have some near neighbours in the realist camp.

In fact, however, I'm a little more optimistic. I think it's possible that by thinking about the kind of world that would 'look like quantum mechanics' from the perspective of creatures embedded in spacetime in the way that we are, we have some sort of handle on the project of constructing a less perspectival successor theory to quantum mechanics. The theoretical stance needed here is the one provided by meta-modelling: a viewpoint sensitive both to the nature of the world and to the nature of the creatures who inhabit that world, in order to address the question as to why *those* creatures should model *that* world in *this* way—i.e., the way exemplified by quantum mechanics, as we have it today.[11]

For me, then, the appeal to quantum probabilities in support of objectivism cuts neither ice nor mustard. As I've already noted, I think the appeal of objectivism about probability in science rests, at best, on a rather blinkered view of the options. In a more extended treatment of these issues, I'd back up that assessment by appealing to more local arguments for subjectivism about quantum probabilities. For example, Jenann Ismael [1996] argues that the common assumption that quantum probabilities are intrinsic physical properties is incompatible with many standard views of the nature of chance—in particular, those views which allow that the chance of an event at a time supervenes on the history of the world at later times. If we give up supervenience to avoid this conflict, we inflate our metaphysics at the cost of our epistemology. In effect, we worsen the problems noted earlier of explaining and justifying upward and downward inference to chances. (As Ismael [1996, p. 89] puts it, 'the injection of metaphysics ... goes against empiricist inclinations'.)

Ismael goes on to suggest that an injection of subjectivism offers a much better prognosis. I agree, and I think the point generalises. In general, perspectival mod-

---

unobservable objects and the external relations between them, rather this structure is ontologically basic. (2003, 45–46, my emphasis)

[11]When I'm very optimistic, I think that the answer might point to a theory less puzzling than quantum mechanics itself, from a realist's point of view. If so, then the satisfying moral would be that excessive realism about modal notions had actually stood in the way of a satisfactory realist interpretation of quantum phenomena.

els of probability—which provide a better way of introducing subjectivity than the self-descriptive subjectivist models we considered in Section 2—make easy work of things that are difficult for objectivist approaches. In particular, they avoid the epistemological and pragmatic problems mentioned earlier—How can we know about chances, and why do they have the relevance they do for decision? And they do well in explaining other oddities of chance, such as its apparent time-asymmetry.

What happens to the analogue of Russell's claim for the case of probability, if we accept this perspectival account of the probabilities in quantum theory and elsewhere? Two essential points. First, Russell is right in thinking that there are no chances objectively construed. But second, he is wrong to think that talk of probabilities only passes muster because we fail to see what is wrong with it. What he misses is the republican option. Probability passes muster as a notion we ourselves bring to our models of the world—a possibility that requires that we see ourselves as Kant does, as conceptual creators, not as mere consumers of conceptual categories pre-packaged by God.

6   Modelling Modalisers

I said at the beginning that I wanted to use probability as guide to the general case, including causation. I don't have space here to try to provide details, but it should be clear how the program goes. We aim for a meta-model of a modal agent— a creature embedded in time, from whose epistemological and decision-theoretic perspective, modal talk makes sense.

In the case of causation, as I noted at the beginning, an attractive idea is that the agent's perspective is crucial. This idea has appealed to many people, one of the first of them Ramsey. As Ramsey puts it, 'from the situation when we are deliberating seems to ... arise the general difference of cause and effect.' (1929, 146) In order to make good this idea, our task as meta-modellers is to show how modelling the world in causal terms serves the interests of creatures we model as abstract agents, embedded in a non-causal environment. (In other words, we don't include causation in the model environment of the creatures represented in our meta-model, but the meta-model aims to show how it is useful for these creatures to include it in their models.)

Two further comments. First, as I've argued elsewhere (Price [1991; 1992a; 1992b; 1996; 2001]), one of the striking advantages of this perspectival approach to causation is that it offers a coherent account of the origins of the asymmetry and temporal orientation of causation. Roughly, it is a projection of our own temporal asymmetry as agents—again, this is something we could use our meta-models to make more precise. As in the probability case, abandoning objectivism for perspectival thus offers light work in place of heavy weather.

In fairness to Russell, it should be noted that he himself is clearly aware of some of the relevant considerations here. In discussing the origins of our intuition that the past is fixed and the future open, he says that 'there seems no doubt that the main difference in our feelings arises from the accidental fact that the past but not the future can be known by memory', [Russell, 1913, p. 203]. We might quibble about the way Russell expresses this. Roughly, it is only accidental that the past and not the future can be known by memory if we interpret the terms as rigid designators, picking out the same temporal direction in worlds in which our own physical orientation is reversed. It is plausible that in that imagined case, we would still use the term 'past' to pick out the direction we remembered. But that semantic point aside, Russell is evidently aware of one of the key contingencies about us, which seems bound to play centre-stage in an explanation of the modal perspective.

The second comment is that there is no reason to be troubled by the fact that in conducting these investigations—in thinking about the usefulness of causal and modal notions for creatures like us—we are employing those very modal notions. For example, we are thinking about how our lives *would be,* if we didn't employ counterfactual reasoning. But this circularity is surely not vicious. On the contrary, it simply serves to confirm that there is little or no interesting science that is free of perspective, in the deep sense associated with the modal categories. If it is science itself that reveals this to us, how are objectivists to respond? By claiming that in this respect science is not to be trusted, and that philosophy offers deeper insights?

7 Why I am not a Fictionalist

I noted earlier that although the perspectival viewpoint could usefully be compared to fictionalism, I wanted to distinguish my own version of the view from fiction-alism. The easiest way to explain this point is to explain the analogous distinction between my view and orthodox noncognitivism or quasirealism [Blackburn, 1993], and then transpose.

A noncognitivist or quasirealist typically makes two claims about the functions of the target discourse—evaluative language, for example. One claim is positive, the other negative. The negative claim says that the terms or statements charac-teristic of the discourse lack some semantic property. They are non-referential, non-truthconditional, non-descriptive, non-factual, or something of the kind. The positive claim offers an alternative account of their functions of the language in question—for example, that it expresses, or projects from, evaluative attitudes. Thus we might say that the negative claim is *anti-representational,* the positive claim *expressivist.*

My version of expressivism keeps the positive claim and throws away the neg-ative claim. As I have argued elsewhere [Price, 2004], this combination seems obligatory if we are deflationists about the relevant semantic notions. For to be a semantic deflationist is to believe, inter alia, that semantic notions play no substantial theoretical role. If that's true, such notions can't play a substantial role in the characterisation of a philosophical position, and the negative claim must be empty of theoretical content. Contrary to received wisdom, then, semantic minimalism is a friend and not an enemy of expressivism.

In other words, ignoring a few bells and whistles, semantic deflationism commits us to a kind of global expressivism. Why? Because without substantial semantic notions to underpin representationalism—to underpin the claim that some discourses are 'genuinely' realist, or factual, or cognitive, or whatever—all that's left to us as linguistic theorists are non-representational accounts of the functions of those discourses. What's left to us, in other words, are the kinds of accounts that noncognitivists and quasirealists have offered locally, in telling the positive part of their story. (Again, all of this could be cast in the terms of meta-models. There, the point is that if representation is not a substantial theoretical notion, perspectival meta-models are all that we have left.)

So am I a quasirealist? Yes, in one sense, for I endorse the quasirealist's project of explaining why expressively-grounded talk takes what are usually seen as realist forms—why it is treated as truth-apt, for example. But no, in another sense, for in endorsing this project globally, I reject the view that there is any bit of declarative language for which this is not the appropriate theoretical perspective. Hence, in effect, in claiming that there is nothing else, I reject the basis for the label 'quasi'.

In one of his late papers [Lewis, 2004], David Lewis argues that quasirealism is essentially fictionalism. I agree, more or less. If there is a difference, it is that in the case of fictionalism the negative claim is more naturally cast in ontological rather than semantic terms. Instead of saying with the quasirealist that evaluative talk is non-referential, the fictionalist may say simply that there are no values. But in this case I distinguish myself from fictionalism just as before, except that the necessary deflationism is ontological, rather than semantic. I think that if, with Carnap [1952], we reject the idea of an ontological standpoint external to our linguistic practices, then there is no viewpoint from which we can properly say that there are no values. Of course, the Carnapian point cuts the other way, too. There is no distinctively philosophical viewpoint from which we can say that there are values. What we are left with is a kind of internal, minimal realism, combined once again with a global expressivist standpoint (which is where all the interesting theoretical work takes place).

Note that Carnap does not require that we reject the fictional–non-fictional distinction altogether. On the contrary, as one of his own examples illustrates. Discussing what he calls 'the thing world'—'the spatio-temporally ordered system of observable things and events' [Carnap, 1952, p. 210]—he says:

> Once we have accepted this thing-language and thereby the frame-
> work of things, we can raise and answer internal questions, e.g., 'Is
> there a white piece of paper on my desk', 'Are unicorns ... real or

merely imaginary', and the like. These questions are answered by empirical investigations. [Carnap, 1952, p. 210]

So Carnap is happy to allow us to distinguish fictional unicorns from real wombats or extinct Tasmanian tigers. However, he stresses that 'from these [legitimate, internal ontological] questions we must distinguish the external question of the reality of the thing world itself'. He notes that this is the kind of question metaphysicians take themselves to be addressing, but argues that

> it cannot be solved because it is framed in the wrong way. To be real in the scientific sense means to be an element of the framework; hence this concept cannot meaningfully be applied to the framework itself ... The thesis of the reality of the thing world ... cannot be formulated in the thing language or, it seems, in any other theoretical language. [Carnap, 1952, pp. 210–211]

Among the metaphysical positions thus disallowed is a *global* fictionalism about the entities of the thing-world. And similarly in the modal case, it seems to me: Carnap's view is quite compatible with *local* kinds of fictionalism about causation, chance, and the like. It allows us to say that the causes and probabilities that occupy Sherlock Holmes' attention are as fictional as the man himself. What is excluded is global fictionalism about the entire causal or probabilistic framework, of the kind advocated by those who call themselves fictionalists about causes, chance and modality.

Within the philosophy of science, a useful comparison for the resulting position is with Arthur Fine's deflationary realism, encapsulated in what he calls the Natural Ontological Attitude [Fine, 1984]. With one crucial qualification, Fine's souped-down realism suits my modal perspectivalist very well. The qualification is that there is one respect in which my view is very much less quietist than Fine's— one respect in which it envisages an important role for philosophy, albeit not the traditional metaphysical role. This concerns the recognition that different models may do very different work for us, being projections from very different contingent grounds. But the standpoint from which we draw such distinctions is not that of a traditional ontologist or metaphysician (opponents I share with Fine). It is that of the psychologist, the linguistic and the anthropologist, reflecting on aspects of human thought and behaviour.

Summing up, then, these are my reasons for distinguishing my view from fictionalism. Like non-cognitivism, fictionalism makes two claims about the function of the talk in question, a positive claim and a negative claim. I reject the negative claim—whether couched in semantic or ontological terms—on deflationist grounds. Deflationism thus leaves me without the theoretical vocabulary or standpoint required to say that the talk is question is fictional. What's left, by default, is a kind of minimal realism—though a pluralistic realism, which insists that different models may do different kinds of jobs.

8   Meta-models and Metaphysics?

In distinguishing my view from fictionalism, I've relied on a deflationary attitude to metaphysics. But it might be argued that meta-models provide what we need to reinflate things—to draw the distinction between cases in which realism is appropriate and cases in which it is not. Haven't I myself suggested as much, in effect, by noting that in the case of modality, we don't need to put modal facts in the modelled environment, in order to explain the modal talk of creatures modelled in our meta-models of modallisers?

This is an important objection, which needs more detailed treatment than I can give it here. Briefly, I think that although we can usefully highlight degrees of perspectivalism in this way, there is no non-perspectival view at the end of the tunnel. The most persuasive argument for this conclusion, in my view, is that the phenomenon of generality in language always supplies an ingredient which we don't need in the modelled environment, in order to account for linguistic practice. Models and theories are always tools we plan to apply in new cases. That's why we have them, as implements with which better to face an unknown future. At any given stage, however, we don't need those unknown cases, in order to account for the models, concepts or theories we currently possess. *All* models, concepts and theories thus reach beyond what we need in the environment, to explain the fact that we possess them. And nothing counts as genuinely representational, if this is our test.[12]

It is worth noting that even if the proposed test did succeed in distinguishing a class of non-perspectival models, it is far from clear that the result would necessarily count as realism. Another possibility, apparently, would be a kind of two-dimensional instrumentalism. Imagine that all scientific language is construed instrumentally, in something like van Fraassen's [1980] sense. That is, the only relevant epistemic attitude is what van Fraassen calls acceptance, not belief. (Perhaps we have decided that there are no genuine beliefs, in van Fraassen's sense, on the grounds that we are semantic deflationists.) Within this instrumentalism, there might then be a place for an additional, orthogonal dimension in which models and theories could count as instruments—the dimension associated with perspectivalism (and the idea that particular conceptual instruments have specialised functions, related to particular needs and capacities). The upshot would be that even if some model and theories were entirely non-perspectival, and hence had

---

[12]This argument is closely related to the claim that the Wittgensteinian rule-following considerations provide an argument for global non-factualism—that is, they reveal a dimension of perspectival contingency which is ineliminable from language. See Price [1988; 1998] and Pettit [1991] for views of this kind.

no second-dimensional instrumental role, they would remain instrumental in the original sense. Realism would remain out of reach.

9 Conclusion

I've suggested that the vocabulary of theoretical models provides a helpful vehicle for pragmatist approaches to the modal notions in science, as to other topics. The value of models is that they tend naturally to bring the foreground the issues of function, use and role in practice, on which pragmatists want to focus—and at the same time (what's really the other side of the same coin) downplay the naïve representationalism that pragmatists see as such an impediment to good philosophy.

Using this model vocabulary, I've tried to delineate the kind of perspectivalism I take to be particularly promising in the modal case, and perhaps especially in the causal case. It is this kind of perspectivalism that I take to provide the best template for an account of causation in terms of agency and manipulation. While I haven't said enough, probably, to convince the two groups of opponents I mentioned at the beginning—those who agree about the importance of manipulation but want to think of it in more objectivist terms, and those who disagree about manipulation altogether—I hope I have made it clearer what the argument is all about. I hope I've also established that the debate about causation connects with much broader issues, in several directions—about other species of modality employed in science, for example, and about the range of options for a philosophical account of any such notion. In particular, while I may not have converted anyone to causal republicanism, I do think I've shown that it belongs on the philosophical map.

Acknowledgements

This material has been much improved by discussions of earlier versions at conferences in Sydney and London, and with a Philosophy of Science group at UNC, Chapel Hill. I am particularly indebted to Jenann Ismael, Raffaella Campaner, Marc Lange and John Roberts. I am also grateful to Mauricio Suárez for many helpful comments, and to the Australian Research Council, for research support.

BIBLIOGRAPHY

[Blackburn, 1993] S. Blackburn. *Essays in Quasi-Realism,* New York: Oxford University Press, 1993.

[Carnap, 1952] R. Carnap. Empiricism, Semantics and Ontology. In L. Linsky, ed., *Semantics and the Philosophy of Language,* pp. 208–228. Urbana: University of Illinois Press, 1952. Originally published in *Revue Internationale de Philosophie*, **4**, 20–40.

[Fine, 1984] A. Fine. The Natural Ontological Attitude. In J. Leplin, ed., *Scientific Realism*, pp. 83–107. Berkeley: University of California Press, 1984.

[French, 1998] S. French. On the Withering Away of Physical Objects. In E. Castellani, ed., *Interpreting Bodies: Classical and Quantum Objects in Modern Physics*, pp. 93–113. Princeton University Press, 1998.

[French and Ladyman, 2003] S. French and J. Ladyman. Remodelling Structural Realism: Quantum Physics and the Metaphysics of Structure, *Synthese*, **136**, 31–56, 2003.

[Hume, 1998] D. Hume. *An Enquiry Concerning the Principles of Morals*, T. L. Beauchamp, ed., Oxford: Oxford University Press, 1998.

[Ismael, 1996] J. Ismael. What Chances Could Not Be. *British Journal for the Philosophy of Science*, **47**, 76–91, 1996.

[Ladyman, 1998] J. Ladyman. What is Structural Realism? *Studies in the History and Philosophy of Science*, **29**, 409–424, 1998.

[Langton, 1998] R. Langton. *Kantian Humility—Our Ignorance of Things in Themselves*, Oxford: Oxford University Press, 1998.

[Lewis, 1980] D. Lewis. A Subjectivist's Guide to Objective Chance. In Richard C. Jeffrey, ed., *Studies in Inductive Logic and Probability*, vol. II. Berkeley: University of California Press, 1980.

[Lewis, 1986] D. Lewis. *Philosophical Papers, Vol. II*, Oxford: Oxford University Press, 1986.

[Lewis, 2004] D. Lewis. Quasi-realism as Fictionalism. In Mark Kalderon, ed., *Fictionalist Approaches to Metaphysics*, New York: Oxford University Press, 2004.

[Lewis, forthcoming] D. Lewis. Ramseyan Humility, forthcoming.

[Menzies and Price, 1993] P. Menzies and H. Price. Causation as a Secondary Quality. *British Journal for the Philosophy of Science*, **44**, 187–203, 1993.

[Morgan and Morrison, 1999] M. S. Morgan and M. Morrison, eds. *Models as Mediators: Perspectives on Natural and Social Science*, Cambridge: Cambridge University Press, 1999.

[Pearl, 2000] J. Pearl. *Causality: Models, Reasoning, and Inference*, Cambridge: Cambridge University Press, 2000.

[Pettit, 1991] P. Pettit. Realism and Response-Dependence. *Mind*, **100**, 587–626, 1991.

[Price, 1988] H. Price. *Facts and the Function of Truth*, Oxford: Basil Blackwell, 1988.

[Price, 1991] H. Price. Agency and Probabilistic Causality. *British Journal for the Philosophy of Science*, **42**, 157–176, 1991.

[Price, 1992a] H. Price. Agency and Causal Asymmetry. *Mind*, **101**, 501–520, 1992.

[Price, 1992b] H. Price. The Direction of Causation: Ramsey's Ultimate Contingency. In David Hull, Micky Forbes and Kathleen Okruhlik, eds., *PSA 1992: Vol. 2*, pp. 253–267. East Lansing MI: Philosophy of Science Association, 1992.

[Price, 1996] H. Price. *Time's Arrow and Archimedes' Point*, New York: Oxford University Press, 1996.

[Price, 1998] H. Price. Two Paths to Pragmatism II. In Casati, R. and Tappolet, C., eds., *European Review of Philosophy*, **3**, 109–47, 1998.

[Price, 2001] H. Price. Causation in the Special Sciences: the Case for Pragmatism. In Domenico Costantini, Maria Carla Galavotti and Patrick Suppes, eds., *Stochastic Causality*, pp. 103–120. Stanford: CSLI Publications, 2001.

[Price, 2004] H. Price. Immodesty Without Mirrors—Making Sense of Wittgenstein's Linguistic Pluralism. In Max Kölbel and Bernhard Weiss, eds., *Wittgenstein's Lasting Significance*, pp. 79–205. London: Routledge & Kegan Paul, 2004.

[Ramsey, 1929] F. P. Ramsey. General Propositions and Causality. In D. H. Mellor, ed., *Foundations: Essays in Philosophy, Logic, Mathematics and Economics*, pp. 133–151. London: Routledge and Kegan Paul, 1929.

[Russell, 1913] B. Russell. On the Notion of Cause, *Proceedings of the Aristotelian Society*, **13**, 1–26, 1913. Reprinted in *Mysticism and Logic*, pp. 171–196, Doubleday, 1953. (Page references are to the latter version.)

[Russell, 1914] B. Russell. The Relation of Sense-Data to Physics. In *Mysticism and Logic*, Doubleday, 1953.

[Strevens, 2003] M. Strevens *Bigger than Chaos*, Cambridge, Mass.: Harvard University Press, 2003.

[van Fraassen, 1980] B. van Fraassen. *The Scientific Image*. Oxford: Oxford University Press, 1980.

[van Fraassen, 1989] B. van Fraassen. *Laws and Symmetry*. Oxford: Oxford University Press, 1989.

[Woodward, 2001] J. Woodward. Causation and Manipulability. In Edward N. Zalta, ed.,*The Stanford Encyclopedia of Philosophy*, 2001.

[Worrall, 1989] J. Worrall. Structural Realism: The Best of Both Worlds? *Dialectica*, **43**, 99–124, 1989.

[Zahar, 1996] E. G. Zahar. Poincaré's Structural Realism and his Logic of Discovery. In Jean-Louis Greffe, Gerhard Heinzmann and Kuno Lorenz, eds., *Henri Poincaré: Science and Philosophy*, pp. 45–68. Berlin: Akademie Verlag & A. Blanchard, 1996.

Huw Price

Centre for Time
Department of Philosophy
Main Quad, A14
The University of Sydney
NSW 2006
Australia
Email: huw@mail.usyd.edu.au

Republican Perspectivalism
A Comment on Huw Price

RAFFAELLA CAMPANER

The main aim of the paper is advancing a broad perspectivalist programme, regarding causation as well as various other modal notions, such as counterfactual conditionals and, especially, probability. Recalling Russell's famous statement about causation, defined as a 'relic of a bygone age', Price puts forward what he labels a 'republican' approach: we shall oppose the 'monarchic' option of believing that our conceptual framework is provided by a 'higher authority', namely imposed by the world itself, and consider, instead, how much of it comes from us. According to the alternative advanced, the modal notions we use in dealing with the world are actually constructed, or, better, 'projected' by us. In a democratic republic citizens happen to be involved in the exertion of political power by choosing their representatives. Analogously—it is suggested—we are actively involved in the 'elections' of modal notions, which are aimed at helping us out in dealing with phenomena. This form of 'perspectivalism', or 'projectivism', is pursued by introducing 'meta-models', namely 'models of the users of models'.[1] In this comment I shall call attention to some problematic aspects this approach might raise and to some further developments it might require.

The projectivist metaphor has a long philosophical history. According to Mark Johnston, it is a passage by Descartes[2] that constitutes the '*locus classicus* of the accusation of projective error'. Basic cognitive and psychological structures, in fact, orient the subjects towards certain categories, but also lead them towards the presumption of the ontological existence of such categories: causal reasoning leads us to maintain that physical causal nexus exist; our credence-based psychology lead us to claim that objective probabilities do, and so forth. One of the main claims of various projectivist positions is that what we mistake for genuine, mind-independent features of our environment and of things themselves are just features of mental responses, manifestations of our experience, actions, forms of receptivity. We seem to get things easily wrong and our projections seem to be error-

---

[1] Price, this volume, p. 51.

[2] According to Descartes, the first judgements which we 'make in our youth, and later in the common philosophy (...) have accustomed us to attribute to bodies many things which pertain only to the soul' (Descartes, *Letter to Abbé de Launey*, 22 July 1641, quoted in [Johnston, 1998, p. 6].

Laws and Models in Science, 73–78.
© 2004, the author.

infected. According to an effective metaphor by Johnston, the result of philosophical projectivism 'is that we are depicted as led on through our environment by the siren song of voices which turn out to be our own'.[3] A merit of Price's proposal consists hence in making our projections the cornerstone of a global philosophical reorientation. Projections are not conceived of as 'a repertoire of elaborate errors',[4] but rather as our mode of access to the world, which is maintained to have always a perspectival face since 'nothing counts as genuinely representational'.[5] Focusing on meta-modelling, we get a correct understanding of projectivism, we obviate to the problems traditionally related to uncritical projections of our mental contents and responses, and realise that 'there is little or no interesting science that is free of perspective in the deep sense associated with the modal categories'.[6]

A first puzzle arises, though, as to whether such an analysis in terms of users of models is easy to achieve. As we have mentioned, according to Price, modal notions are to be understood by reflecting upon our needs, characteristics and aims in modelling them. In the case of causation, what are at stake is our being creatures embedded in time and our acting in the present, on the light of past, for future ends: we intervene on the world and manipulate some event in order to obtain a given effect. According to the definition given in the 'agency theory', 'an event A is a cause of a distinct event B just in case bringing about the occurrence of A would be an effective means by which a free agent could bring about the occurrence of B'.[7] Something analogous holds in the case of probability, where the crucial perspective is that of creatures endowed with a credence-based psychology. When we focus on meta-modelling, though, it seems we are required to conceive of some 'abstract agents', embedded in a 'non-causal', and 'non-probabilistic' or, more generally, 'non-modal', environment.[8] It looks at least problematic for us to be able to step back from our being involved in agency, our being embedded in time and our having credences, given these features are so deeply entrenched with our being human creatures. Furthermore, since these seem all to be fundamental characteristics of us as human beings, we might wonder what we are actually left with under the label of 'abstract agent'. Yet, only a point of view—so to speak—'from outside ourselves', only a partial abandoning of our distinctive human constitution might give us awareness about being constantly inside a perspectival game, and allow us not to be just like passive observers, like passive subjects to an absolute king with no chance to exercise any political power whatsoever.

A second, related issue concerns the very notion of 'distinctively human perspective'. Price's perspectivalism evokes Kant and the Copernican revolution he

---

[3][Johnston, 1998, p.5].
[4][Johnston, 1998, p. 4].
[5]Price, this volume, p. 66.
[6]Price, this volume, p. 63.
[7][Menzies and Price, 1993, p. 187]. See also Price [1991; 1992; 1993].
[8]Price, this volume, p. 62.

promoted, which provoked an extremely deep change in our view of the origins of some categories. Price's own republican option requires that 'we see ourselves as Kant does, as conceptual creators, not as mere consumers of conceptual categories pre-packaged by God'.[9] The idea of some Kantian 'creative' conceptual power is thus inserted into the meta-modelling discourse, with its emphasis on model-users. At the same time, Price's perspectivalism is complementary to his pragmatism, in another paper of his defined as involving 'practice-subjectivity', namely as making 'central reference to the role of concepts in the lives and practice of creatures who use them'.[10] In presenting his broader perspectivalism, Price points out that a discourse on meta-models has to be concerned with the utility of models for model-users. He highlights that in order to address such an issue 'we need to include enough 'environment' in the model, and enough detail concerning the users themselves'[11] and we shall 'consider the relevance of variations in characteristics of the users'.[12] Models are regarded here as valuable because of their role and use *in practice*. That is why we shall be interested in changing aspects, raising 'the question as to why *those* creatures should model *that* world in *this* way'.[13] Price's perspectivalism is grounded on features that characterise our own constitution, and hence access to the world, on a permanent basis, as suggested by Kant. At the same time, though, modal notions and the universal features we are endowed with are here to be seen in relation with specific circumstances, in the context of richly textured and varying situations. It could be further specified how, on the one side, recalling of the Kantian view with respect to our distinctive and definite features and, on the other side, pragmatism can be fit together, how they can be 'attuned' to clarify what we mean by saying that our conceptual tools depend on 'the kind of creatures we are'.[14]

On a more substantial level, we shall notice that the form of perspectivalism presented here, although uneliminable, seems to admit of 'degrees'. On the one hand, Price promotes a *global* perspectivalism, being ready to live with the fact that we could realise we cannot but see everything through, so-to-speak, some 'distinctive spectacles', that we could know nothing not tainted with modal perspectivity. On the other hand, with respect to quantum mechanics he declares he is 'a little more optimistic', believing that 'it is possible that by thinking about the kind of world that would 'look like quantum mechanics' from the perspective of creatures embedded in spacetime in the way that we are, we have some sort of handle on the project of constructing a *less perspectival* successor theory to quantum mechan-

---

[9] Price, this volume, p. 62.
[10] [Price, 2001, p. 106].
[11] Price, this volume, p. 53.
[12] Price. this volume, p. 51.
[13] Price, this volume, p. 61.
[14] Price, this volume, p. 57.

ics'.[15] But how are we supposed to identify and define 'more perspectival' and 'less perspectival' accounts? Would the former be more inclined towards subjectivist positions, and the latter include a more 'objectivist flavour' and prevent the concerns of pragmatism and realism from appearing 'orthogonal'?[16] Price himself proposes his republican option as an alternative between eliminativism and full-blooded realism.[17] Shall we think that such a position can be, in turn, articulated into further kinds of perspectivalism? What we project are held to be features of the world *'as it looks' to us from our distinctive standpoint*.[18] How can we be 'more' or 'less' situated in such a distinctive standpoint? Perhaps it should be clarified in which sense our 'perspectival look' can admit of different 'degrees'.

Price's proposal focuses on models and their use for creatures like us. His republican view originates from a reconsideration of specific human structures, structures of believers or decision-makers in the case of probability, and structures of agents in the case of causation. Finally, then, a natural question arises as to whether the other modal notions mentioned by Price depend on *other* human features? Since it is claimed that 'there is no non-perspectival view at the end of the tunnel',[19] how many other aspects of our cognitive structure or epistemic situation are involved, and what can this perspectivalism precisely be extended to? Does it apply to all scientific knowledge? Does it cover non-scientific knowledge as well? Price states that 'Russell seems largely blind to the republican possibility'.[20] In referring to the subjectivist turn in modern philosophy, from Descartes onwards, Russell points to the risk of an *anarchic* outcome: 'with subjectivism in philosophy, anarchism in politics goes hand in hand'.[21] Even if subjectivism and perspectivalism are distinct positions, here too an extension of the perspectival approach to *all* the notions we use might make us fall from the monarchic option into an anarchic regime. We have been presented with 'causal republicanism' and 'probabilistic republicanism';[22] we can imagine along the same line of thought a 'counterfactual-republicanism', and we would need more details about how to develop further 'modal-republicanisms', as suggested. Given it is maintained that knowledge is perspectival all way down, there could be a huge, potentially infinite number of modelled notions, and hence a potentially infinite number of perspectives, unless we clearly determine what the relevant ones are.

As Price recalls in 'Two Paths to Pragmatism II', the most obvious cases of concepts which exhibit a form of dependence from human features are the tradi-

---

[15] Price, this volume, p. 61.

[16] Price, this volume, p. 51.

[17] See Price, this volume, p. 50.

[18] Price, this volume, p.53. See also p. 58.

[19] Price, this volume, p. 66.

[20] Price, this volume, p. 50.

[21] [Russell, 1979, p. 20, italics added. See also p. 481].

[22] Price, this volume, p. 50 and p. 59.

tional secondary qualities. There the matter seems to be much easier, given that such dependence rests on perceptual abilities and on the functioning of sensory means, which are limited and basically uniform through time. Matching a conceptual framework and human perspectives will be a much more complex operation, open to a much wider range of solutions. To avoid falling into 'anarchic projections' from *any* contingent perspectives we might hold, we believe a restricted list of the fundamental, privileged standpoints should be identified. We would then get a kind of multi-faceted, pragmatic and genuinely democratic perspectivalism, into which only a limited number of 'parties' shall be admitted. For perspectivalism to be effectively generalized 'via models with a more human face',[23] we need to specify which are all the human perspectives we are taking into account, and what they are exactly like. Price maintains that 'clarifying the perspectival option is a matter of locating it on philosophical maps'.[24] We believe that we should also make, in his own terms, some more epistemological 'cartography'[25] with regard to all the 'key contingencies about us'[26] on which this 'republican perspectivalism' rests.

BIBLIOGRAPHY

[Halpin, 2003] J.F. Halpin. Scientific Laws: A Perspectival Account. *Erkenntnis*, **58**, 137–168, 2003.
[Hausman, 1998] D. Hausman. *Causal Asymmetries*. Cambridge: Cambridge University Press, ch. 5, 86–110, 1998.
[Johnston, 1998] M. Johnston. Are Manifest Qualities Response-Dependent? *The Monist*, **81**, 3–43, 1998.
[Menzies and Price, 1993] P. Menzies and H. Price. Causation as a Secondary Quality. *British Journal for the Philosophy of Science*, **44**, 187–203.
[Pettit, 1991] P. Pettit. Realism and Response-Dependence. *Mind*, **100**, 587–626, 1991.
[Price, 1991] H. Price. Agency and Probabilistic Causality. *British Journal for the Philosophy of Science*, **42**, 157–176, 1991.
[Price, 1992] H. Price. Agency and Causal Asymmetry. *Mind*, **101**, 501–519, 1992.
[Price, 1993] H. Price. The Direction of Causation: Ramsey's Ultimate Contingency. In D. Hull, M. Forbes and K. Okruhlik (eds.), *PSA 1992* II, East Lansing, Michigan: Philosophy of Science Association, 253–267, 1993.
[Price, 1996] H. Price. *Time's Arrow and Archimedes Point*. New York: Oxford University Press, 1996.
[Price, 1998] H. Price. Two Paths to Pragmatism II. In R. Casati and C. Tappolet (eds.), *Response-dependence, Euroepan Review of Philosophy*, **3**, 109–147, 1998.
[Price, 2001] H. Price. Causation in the Special Sciences: the Case for Pragmatism. In M.C. Galavotti, P. Suppes and D. Costantini (eds.), *Stochastic Causality*, Stanford: Csli Publications, 103–120, 2001.
[Price, 2004] H. Price. Models and Modals. In D. Gillies (eds.), *Laws and Models in Science*, pp. 49–69. London: King's College Publications, 2004.
[Russell, 1913] B. Russell. On the Notion of Cause. *Proceedings of the Aristotelian Society*, **13**, 1–26, 1913.
[Russell, 1979] B. Russell. *History of Western Philosophy*. London: Unwin Paperbacks, 1979.

---

[23] Price, this volume, p. 51.

[24] Price, this volume, p. 49.

[25] Price [Price, 2001, p. 104].

[26] Price, this volume, p. 63. See also p. 55 and p. 57.

[Sainsbury, 1998] R.M. Sainsbury. Projections and Relations. *The Monist*, **81**, 133–160, 1998.
[Woodward, 2001] J. Woodward. Causation and Manipulability. *Stanford Encyclopedia of Philosophy*, 2001, http://plato.stanford.edu/entries/causation-mani/

Rafaella Campaner

Department of Philosophy
University of Bologna
Via Zamboni 38
40126 Bologna
Italy
Email: campaner@philo.unibo.it

Truth and the Aim of Belief

PASCAL ENGEL

1 Introduction

It is often said that 'belief aims at truth'. This is presented sometimes as a truism, sometimes as capturing an essential and constitutive feature of belief and of inquiry. It is a truism that to believe something is to believe that it is true, and that in this sense beliefs are directed towards truth. It is also a truism that we aim at having true beliefs rather than false ones, and that in this sense truth is our goal when we form beliefs. But it is false if it is supposed to apply literally to beliefs rather than to believers: obviously beliefs do not 'aim' at anything by themselves, they do not contain little archers trying to hit the target of truth with their arrows. In this sense the claim must be metaphorical. It is more appropriate, if there is an aim at all, to ascribe it to persons and not to beliefs. But in this sense too it is false: sometimes we do not aim at having true beliefs, but at having pleasurable or comforting ones. Or perhaps what the claim says is that we should aim at having true beliefs. So what does it mean to say that beliefs aim at truth? And is it true?

There are least three reasons to be interested in clarifying what, to use a generic term, the truth-directedness of belief[1] means. The first is that it promises to tell us what belief is, as a mental state, and what distinguishes it from other kinds of mental states. In particular it is held that truth-directedness is the feature which differentiates belief as a cognitive mental state from motivational states, such as desires and wants. Indeed it is also said to be what prevents belief from being subject to the control of the will.

The second reason is that the truth-directedness of belief seems to have something to do with another feature of belief, its 'normative' character. Beliefs are correct or incorrect, rational or irrational, justified or unjustified. And the fact that beliefs aim at truth, in the sense that truth is the fundamental dimension of assessment of beliefs seems to be related to this normative dimension. So this normative sense should shed light on the nature of our epistemic norms and principles.

The third reason has to do with the nature of truth as a goal or as the main theoretical value. This too is often presented as a truism: truth is the ultimate

---

[1] I borrow this phrase from Velleman [2000]. Other writers, for instance Walker [2001], talk about 'truth aimedness'

Laws and Models in in Science, 79–99.
© *2004, the author.*

epistemic aim of scientific inquiry. But is it really a truism? Some pragmatists and most postmodernists disagree, but also some instrumentalist philosophers of science, who say that the goal of scientific inquiry is not truth but empirical adequacy. So it promises to illuminate some of the fundamental issues about scientific realism and about the value of knowledge in general.

It would be completely unrealistic to hope to deal with all these issues here. I set myself a more modest objective, which is to try to investigate the relationships between these various dimensions of truth-directedness. There is actually a tempting picture of these relationships, which promises to unify these various dimensions. If truth is the ultimate goal or norm of scientific inquiry, that at which our beliefs are directed, then it is plausible to claim that belief is the kind of attitude through which we try to reach this goal. Hence to believe something is not simply to entertain a proposition which happens to be true or false, but it is having a relation to this proposition with the aim that it is true and of avoiding error. The connection seems to be this: if a form of directedness or of direction towards a goal—i.e something purposive—is essential to the nature of belief, we might hope that this very feature is what accounts for the fact—if it is a fact—that truth is essentially a norm and a value. And one might hope to understand this connection by discovering in what sense belief is an essentially purposive activity, which involves, on the part of the believer, a certain kind of intention, hence of action. Truth, in this sense, would not be simply a cognitive or epistemic norm, but a practical one, a goal towards which we direct certain actions. Such a view would help to bridge the gap between theoretical and practical reason. And indeed it is how it is often understood (Velleman [2000], Noordhof [2001]. But I want to argue that this view is false: truth is not a goal of belief in the purposive or in the action sense. It follows the sense in which we can talk of truth as a norm is not this goal oriented sense.

I shall distinguish three main interpretations of the truth directedness of belief, from the weakest to the strongest: causal, normative, and intentional. I shall argue that the causal account is correct, but insufficient to characterise belief properly, and that the intentional account is incorrect. Only the normative sense gives us the right account of the aim of belief. But I shall also argue that the norm which governs belief is better understood as consisting in *knowledge*.

2   The Causal-functional Account of Truth-directedness

The first sense, and *prima facie* the least controversial, in which one may interpret the claim that beliefs aim at truth is a descriptive sense, which tells us something about what beliefs are. It is simply a fact about beliefs, as a particular kind of mental state, that they have contents which are susceptible of being true or false. These contents are usually taken to be propositions, and it is in this sense that beliefs are said to be propositional attitudes. Truth-directedness just expresses this essential feature of beliefs, and in this sense there is no more to the notion of 'aim' than this relationship of a believer with true or false contents. Now of course beliefs can be false. Should we then say that they 'aim' at falsity as much as they 'aim' at truth? No, of course. A proposition, as a possible object of belief, has what Wittgenstein called a 'bipolarity', but a belief, when it has a proposition as its object, is normally directed towards only one pole. Beliefs are candidates to truth, not to falsity. For to believe that P is to believe that P *is true*, and, as Moore's paradox reminds us, it is *prima facie* odd to say: 'I believe that P, but P is false'. This indicates that the normal function of beliefs is to be directed at true propositions rather than false ones. Now it is important that directedness in this sense is not to be understood as implying any form of intention of getting the truth from the believer's part. Beliefs can perform this function without the believer being in any sense conscious that they do. It's just the causal role of beliefs that they are the kind of mental states which register true information. Of course they often fail to perform this function, which is nevertheless their normal function.

There are two familiar ways in which one can cash out the notion truth-directedness in this platudinous descriptive sense. One is to use another familiar metaphor, that of the 'direction of fit' of beliefs [Anscombe, 1958; Searle, 1983]. Beliefs, unlike desires and other motivational states, have a 'world-to-mind' direction of fit: they are supposed to fit the world when their contents are true. Desires have a 'mind-to-world' direction of fit: they are states such that the world is supposed to fit them, and in this sense they are satisfied, but not true. Or, if one prefers, the criterion of success of a belief is truth, whereas the criterion of success of a desire is satisfaction. But the direction-of-fit metaphor does not seem to tell us more than the platitude itself (e.g. [Humberstone, 1992; Sobel and Copp, 2001]).

The other way consists in considering truth-directness as entailed by the familiar dispositional-functionalist conception of belief. According to this conception, an attitude is a belief only if it disposes a subject to behave in certain ways that would tend to realise her desires if the proposition towards which it is directed is true. More precisely, we can formulate the functionalist conception in terms of possible worlds, along the lines proposed by Robert Stalnaker: to believe that P is to have a disposition to act in ways that would tend to satisfy one's desires in a world in which P is true [Stalnaker, 1984, p. 15][2] We can also draw the contrast between beliefs and desires in dispositional terms:

> A belief that P tends to go out of existence in the presence of a perception with the content that not P, whereas a desire that P tends to endure, disposing the subject to bring it about that P. [Smith, 1994, p. 115]

---

[2]For a defense of this view as an interpretation of the aiming at truth feature and against belief-voluntarism, see Bennett [1990].

The same idea is often formulated thus: when one discovers that one among one's beliefs is false, normally one tends to abandon it or to revise it. Many writers have appealed, implicitly or explicitly, to this functionalist platitude to account for the fact that one cannot believe at will. Indeed, it is in this context that Bernard Williams [1970] first coined the phrase 'Beliefs aim at truth'. One of his main points was that to believe at will, irrespective of whether the content of a belief is true, would be impossible, because such 'beliefs' would not have the normal causal role of beliefs, which is to serve as intermediaries between perceptual inputs and behavioural outputs:

> A very central idea with regard to empirical belief is that of coming to believe that P because it is so, the relation between a man's perceptual environment, his perception, and the beliefs that result. Unless a concept satisfies the demands of that notion, namely that we can understand the idea that he comes to believe that P because it is so because his perceptual organs are working, it will not be the concept of empirical belief. [Williams, 1970, p. 149]

Although Williams here talks about the *concept* of belief, the functionalist platitude concerns the *nature* of belief as a mental state. It means that it is the essence of belief to be a disposition to cause certain behaviours given certain desires and such that the proposition believed is true. But the sense in which it is the essence or nature of belief is a *causal* one: as a matter of fact belief is the kind of attitude which has this causal role. It is not because a belief, by definition, is the kind of attitude which is directed towards truth that it has this kind of causal role; on the contrary it is because it has this kind of causal role that it is the kind of attitude which is directed towards truth.

Truth-directedness and functional role are certainly minimal features of belief. But do they suffice to characterise an attitude as belief? No. In the first place, the fact that to believe that P is to believe-true that P, although it can serve to distinguish belief as a propositional attitude from mental states which are not propositional (such as *qualia* or feels) is hardly individuative of belief itself as compared to other propositional attitudes: for instance wishing that P entails wishing that P is true, hoping that P entails hoping that P is true, and even desiring that P—at least when desiring is propositional—entails desiring that P is true. The direction of fit metaphor and the functionalist platitude are supposed to sort out conative attitudes (desires, wishes, wants, volitions, etc.) from cognitive attitudes. But they do nothing to sort out belief from other cognitive attitudes: thinking that P, considering that P, judging that P, seeing that P, learning that P, or even imagining that P also have the 'world to mind' direction of fit. They too, in this sense, can be candidates to truth. It can be held that all these attitudes involve belief in some sense, but the criterion is certainly not enough fine-grained to distinguish belief from other cognitive attitudes [Velleman, 2000, p. 249].

3   The Normative Account of Truth Directedness

It is not enough to say that belief is an attitude which tends causally to produce certain effects in response to certain inputs; nor it is enough to say that a subject who discovers that she has a false belief is disposed to revise it. In the first place, the functionalist platitude leaves out something essential: beliefs are not simply states which instantiate certain causal relations, but also states which enter into rational relations. Beliefs have normative properties, i.e properties such that if certain patterns are not instantiated, they are incorrect or wrong. It is not easy to say what a normative property is, but we can say that normative properties are properties which give rise to certain kinds of *oughts* (without trying here to analyse further the nature of these *oughts*).[3] For instance someone who believes that P, and that if P then Q, *ought* to believe that Q. This is not a matter of a regularity linking the mental states endowed with these propositional contents; it is that if someone has the two first beliefs, but does not have the third, there is something *wrong*. Similarly one ought not to have inconsistent beliefs. So there are things that we ought to believe, and this fact is essential to the nature of belief. Among these are the constraints which weigh upon the evolution of belief under the impact of evidence. Someone who believes that a certain amount of evidence supports an hypothesis ought to believe the hypothesis. Someone who believes that an hypothesis is better, or simpler, than another ought to believe it, etc. The problem here is not to specify the exact nature of these constraints, which is controversial. But it is reasonable to suppose that there is one constraint which is basic for belief: we ought to believe *true* propositions. And we may suggest that this normative property of belief is what we mean when we say that beliefs aim at truth. The aiming-at truth property is here a fundamental dimension of epistemological assessment of our beliefs, the basic norm for belief. Let us call this the *norm of truth (for belief)*.[4] Let us try to spell out what the norm in question is. It cannot be formulated as the imperative to believe any proposition when it is true:

(NT*) *For all P, if P is true, then you ought to believe that P*

For there are plenty of propositions which are true, and which are not worth believing, if only because we we cannot clutter our minds with trivialities. For instance most of the bits of information that I can get from a phone book are true, but no one should bother to believe all these things. Similarly that there are 235451

---

[3] It's a disputed question what the exact form for expressing normative properties is. See in particular Broome [2000]. I adopt here the framework of what I call normative conditionals in Engel[2002]. On these issues, see also Zangwill [1998] and Wedgewood [2002].

[4] It is important to specify that it is a norm for *belief*, since it should not imply that truth is *in itself* a norm. I investigate this issue in [Engel, 2000] and Engel [Engel, 2002, ch. 5].

blades of grass in my garden may be true, but why should I believe it? Likewise, too, it would be absurd to read the symmetric of NT* as forbidding us to believe any falsehood. For it is common experience that we do. *Ought* implies *can*, and it would be crazy to oblige someone to believe *all* truths and to disbelieve *any* falsehood. A more perspicuous formulation of the norm of truth should not say that we ought to believe any truth whatever, but that a belief is correct only if the believed proposition is true. Hence the proper formulation of the norm of truth should not be (NT*), but:

(NT) *For all P, one ought to believe that P only if P*

which does not have the consequence that any true proposition must be believed or is worth believing.

In what sense does (NT) explicitate the truth-directedness of belief? As we have just seen it says that beliefs do not have only causal properties, and that truth-directedness is not simply a causal feature of beliefs: it is a normative property of beliefs, where a normative property is expressed by a conditional of the form (NT). The norm of truth for beliefs seems to be basic or minimal with respect to the other epistemic norms: for someone who believes that P, and that if P then Q, believes that these propositions are *true*, and on that basis, believes that Q is *true*; if he believes that P is *true*, then he must not believe that not P is *true*; and if he believes that evidence E is evidence for the *truth* of hypothesis P, then he ought to believe P, and so on.

Now in what sense is (NT) supposed to be a norm? In general to what extent does the fact that belief has such normative properties individuate belief as a mental state? For instance is it the case that from the fact that beliefs ought to be closed under *modus ponens* that there cannot be someone who believes P, and if P and Q, but does not believe Q? Of course there are such people, at least occasionally. Should we then say that they do not believe P nor if P then Q?[5] A number of beliefs are irrational, inconsistent, and so on, and they do not seem not to beliefs for that. The same objection can be addressed to the norm of truth: can't we aim at having *false* beliefs? Can't we aim at having beliefs for other reasons than the fact that they are true or justified? It is precisely at this point that Williams formulates his famous argument against the possibility of believing 'at will'. If such believing were possible, says Williams, one could have belief irrespective of whether it is true or not, hence irrespective of the basic norm of truth. But, he argues, in that case the subject would be in a contradictory state:

> It is not a contingent fact that I cannot bring it about just like that that I
> believe something, just as it is a contingent fact that that I cannot bring

---

[5] Jackson [2000, p. 101] notes that the normative constraint clashes with Stalnaker's account of belief in terms of possible worlds: subjects who believe that P and that if P then Q, but do not believe Q will be subjects who do not not have a single system of beliefs.

about, just like that, that I am blushing. Why is this? One reason is connected with the characteristic of beliefs that they aim at truth. If I could acquire a belief at will, I could acquire it whether it was true or not; moreover I would know that I could acquire it whether it was true or not. If in full consciousness I could acquire a belief irrespective of its truth, it is unclear that before the event I could seriously think of it as a belief, i.e as something purporting to represent reality. [Williams, 1970, p. 148]

Williams' argument, as an argument against the possibility of believing at will, has been much questioned, and it is not my purpose here to examine it[6]. But it indicates clearly what role the norm of truth is supposed to play in the individuation of a given mental state as a belief. A subject cannot ascribe to herself a belief unless she recognises that belief is constrained by the norm of truth. So the connection seems to be this: if a subject claims to have a belief, in a conscious way, then she cannot fail to recognise the normative requirement (that only truths should be believed). To think of oneself as believing that P involves at least the tacit recognition of this norm, for it would be incoherent, as Williams suggests, to think of oneself as having a belief that P without thereby aiming at having a true belief. In this respect the connection between the very concept of belief and the epistemic norm of truth is conceptual or constitutive.

Several writers have proposed to take the conscious recognition of the norm of truth as the basic criterion for a belief. Thus:

... unless the attitude-holder has what we might call a controlling background intention that his or her attitudinizing is successful only if its propositional content is true, then the attitude is not that of belief. [Humberstone, 1992, p. 73]

To believe a proposition is regarding a proposition as true with the aim of accepting it as true. [Velleman, 2000, p. 251]

Believing, in this sense, is not simply taking a proposition to be true, nor being disposed to act on the basis of its truth; it is endorsing it, and being committed to its truth. As Velleman says, this sorts out belief both from the other propositional attitudes and from the other 'doxastic', or 'belief-like' attitudes such as thinking, considering, or supposing. It also sorts it out from imagining:

What distinguishes believing a proposition from imagining or supposing it is a more narrow and immediate aim—the aim of getting the truth value of that particular proposition right, by regarding the proposition as true only if it really is. [Velleman, 2000, p. 252]

---

[6] See [Winters, 1979] and [Engel, 1999] for further discussion

According to Velleman, this definition of the aim of belief is only minimal, and allows a variety of relationships between a subject and the propositions that he believes to be true. He allows that his definition could apply to unconscious beliefs, or to a sub-system within the agent, and not to conscious beliefs [Velleman, 2000, p. 253]. But the most natural understanding of his definition concerns the case when a person intentionally aims at truth, by forming an act of judgement:

> He entertains a question of the form 'p or not p?' wanting to accept whichever disjunct is true; to that end he accepts one or the other proposition, as indicated by evidence or argument; and he continues to accept it so long as he receives no evidence or argument impugning its truth. the resulting cognition qualifies as a belief because of the intention with which it is formed and subsequently maintained by the believer, and because of the way in which that intention regulates its formation and maintenance. [Velleman, 2000, p. 252]

It is this interpretation of truth directedness which I want to criticise.

4 The Intentional Account of Truth-directedness

Suppose, then, that believing is identified with (a) the conscious recognition of the basic norm of truth and (b) the intention to respect and maintain this norm in the formation of one's beliefs. Let us call this the *intentional* interpretation of truth-directedness. It does not imply that we should believe as many truths as possible. It also allows us to distinguish clearly believing from a disposition to act as if the belief were true, for this disposition can exist without any recognition of the norm of truth on the part of the subject. Finally it incorporates Williams' reason for denying that belief is voluntary: if when I regard a proposition as true, I accept it with the aim of accepting it as true, and if I do so consciously, I cannot at the same time will to believe it while thinking that it is not true, and believe that the reason for which I accept it has nothing to do with its truth. So in this sense, the intentional proposal entails the belief cannot be under voluntary control.

Nevertheless, the very fact that belief is defined as a purposive activity does imply some sort of control. If believing is essentially purposive, belief must have a goal and the goal of a belief has got to be something external to belief itself. For otherwise the mere fact that we take a proposition to be true would achieve this goal. Truth is well suited to be the aim of belief because, in general, the truth of a proposition is quite independent of whether it is believed by me.

On the picture described here, the norms of belief formation are means towards an end which is the goal of believing, in the same sense as that in which certain actions are instrumental in getting a certain result. In this sense, believing implies a form of deliberation, in order to compare our beliefs with the objective of reaching the truth and retaining only those which agree with this objective. So one important consequence of the intentionalist view of truth directedness is that there is much more similarity between belief, as the output of theoretical deliberation, and action, as the output of practical deliberation:

> Reasons for acting would be considerations relevant to the constitutive aim of action, just as reasons for believing are indicators of truth, which is the constitutive aim of belief. and anyone who wasn't susceptible to reasons for acting, because he had no inclination toward the relevant aim, wouldn't be in a position to act, and therefore wouldn't be subject to reasons for acting, just as anyone who has no inclination towards the truth isn't in a position to believe and isn't subject to reasons for belief. [Velleman, 2000, p. 189]

On this view, aiming at truth is the major premiss of a practical reasoning ('Let me possess the truth [about P]' or 'Let us make judgements (about P) for which there is preponderant theoretical reason'), which has a minor premiss ('If I assent to P I will be making a judgement for which I have preponderant theoretical reason'), and which as a judgement as a conclusion, which is a decision. (See [Walker, 1996]). Paul Noordhof describes this option (which is not his own in whis way:

> The aim of truth is not internal to judgement, but applied from the outside via consciousness. Consciousness makes manifest the attractiveness of being disposed to act upon the truth. ... The basic idea is that it is part of the nature of conscious attention that it gives determinative weight to the norm of truth. It makes truth-likelihood the determinate factor in the formation of a certain kind of motivational state. [Noordhof, 2001, p. 258]

The direct consequence of this is that the norm of truth is a practical norm:

> One reason for thinking that the norm of truth is a practical norm is that both intending to judge that p and judging that p are actions. The norm of truth provides considerations for acting in these ways. Broadly conceived, practical norms are precisely those which provide considerations for action. A second reason is that agents act so as to satisfy their desires. An agent's desires are only satisfied as a result of the agent's action if the beliefs and judgements upon which the agent acts are true. Therefore is part of practical reason that beliefs be true. [Noordhof, 2001, p. 263]

But this account is very problematic on several counts.

In the first place it does not allow us to sort out beliefs from other kinds of attitude, such as guesses, suppositions or conjectures.[7] Guesses, like beliefs, are

---

[7] Here I am much indebted to Owens [2003].

true or false, and they have the world to mind direction of fit. The intentional aim
of a guess is truth too. In that respect guesses are quite unlike imaginings, for
instance. In imagining that P (for instance that I am an Oxford don), I imagine
that it is true that I am an Oxford don, but the success of my imagining does
not depend upon the truth of its content in the way in which the success of a belief
does: I may successfully imagine myself wearing a gown, sitting at the High Table,
without it being possibly true that I am in this stance. Similarly I may suppose or
hypothesise that P without it been necessary that I suppose that P be actually
true [Velleman, 2000, pp. 250–252]. By contrast, when we guess we have the
purpose of getting things right. Now if the intentional analysis of aiming at truth
were correct, we should expect that guessing should be governed, like believing,
by epistemic norms, and in the first place by the norm of truth. But clearly guesses
differ from beliefs in the amount of evidence which they require. Of course some
guesses are more serious than others, but the epistemic standards which govern
guess are by definition lower than those which govern beliefs. (The reason for
this is that belief is much more closely related to knowledge than guessing is; see
below Section 4).

In the second place, the intentional account of aiming at truth applies more
readily to states of mind which are admittedly close to belief, but which are distinct
from it, such as judgements and acceptances. To accept that P is to take it from
granted that in a context, without necessarily believing that P. For instance I
may accept that this student is good, in order to encourage him, although I do not
believe it. Acceptance is certainly an intentional act. In this sense, acceptances
doe not aim at truth in the same sense as the one in which beliefs do[8]. 'Judgment,
on many analyses, and notably on Descartes', is indeed a voluntary act of assent
to the truth of a proposition. And indeed such descriptions of believing quoted
above fit perfectly the activity of judging. It fits also the activity of inquiry. In
inquiry, as it traditionally described (e.g. by Descartes, or by Peirce) we do take
into explicit consideration our objective, truth, and we take the steps which are
in accordance with this objective. Inquiry is an intentional and reflective activity
directed to the aim of truth for a given set of beliefs considered during a certain
span of time. It is plausible to say, in this sense, that truth is the goal of inquiry,
and that the inquirer has a reflective attitude towards this goal. Nor does it imply
that the inquirer must believe only truths. He must, as much as he can, but he
should also know that he may let pass a number of falsehoods. It is the point of
engaging in processes of belief revision. In contrast, the believer of an individual
belief does not consider the aims of inquiry when he comes to believe a given
proposition. This is not because the general aims of inquiry do not figure in the
background of believing or that they are irrelevant to it, on the contrary (otherwise

---

[8]There is a large literature on the difference between belief and acceptance. It is reviewed in part in
Engel [2000]

belief would not be subject to norms). It is simply because the believer does not have *to reflect* about the norms or goal of inquiry to be subject to them. On the intentionalist reading my attending to my reasons and my attending to my belief is supposed to show that I am responsible for this belief. But why is this power of reflection supposed to show that I am in control over this belief? Could it be because I attend to the norms of belief, and reflect that the norm 'believe what you take to be true according to the evidence at hand' forces me to entertain the belief? But being aware of the norm does nothing to move me to belief.

A more plausible description of the situation is this. We have first-order beliefs, which are rational or not. We do not have any control over them. They are just forced upon us. Neither do we have control over the norms of rationality. We do not become rational just by being aware of our beliefs and of the norms which govern them. Certainly if a conflict arises between our first-order beliefs and what we believe that she ought to believe, then we should try to change our beliefs. But we do not do that because we is aware of these norms and of our beliefs. We do it because we have to comply by these norms, and are forced to do that.

A weaker version of the intentional account is, however, more plausible. We could say that awareness of our own beliefs and of the norms of reasons is a pre-condition of rational control, and that *if* we have this awareness, then we can be in position to accept certain beliefs and reject others. But it does not say that control is effected *by* this very awareness. It says that it is a necessary, but not a sufficient condition. Certainly to be able to maintain a certain belief, to stick to it, or to reject it, I must be conscious of it. And to assess it, it seems that I must be aware of the norms which govern it[9]. But we should not confuse our second-order beliefs about the way we maintain our beliefs and revise it with the conditions of formation of our first-order beliefs.

A third reason that we have to doubt the coherence of the intentional reading of truth-directedness is that it puts on a par theoretical norms and practical norms. On the intentionalist model, there is a parallel between the prudential or practical reasons for belief on the one hand, and the epistemic or theoretical reasons on the other. Even if one does not raise here the question whether one kind of reasons can be overridden by the other, or reduced to it (see Appendix), there is a strong asymmetry here. We have an intrinsic authority over what we do which cannot be transferred to the kind of authority which is claimed by the intentionalist reading of truth-directedness to apply to belief. In the practical sphere, we can say: 'I am willing to ϕ but I ought not to'. My reasons do not lose their probative force because I have decided to ϕ. But in the theoretical sphere, there is something in 'I believe that P, but I ought not to'. You may be denied any experience of evaluating contrary to reasons, but you are by no means denied the experience of evaluating contrary to evaluation. Failure to exercise my free will are common and

---

[9]I am indebted here to the discussion in Owens [2000] and [2003].

recognisable. But failures to comply by the norms of theoretical reason are elusive. (This is why it is in general easier to be conscious of being akratic than it is to be conscious of being a fool). In other words, my judgements about what I ought to believe lack the intrinsic authority over my beliefs which my practical judgements enjoy about my actions.[10] In this sense, the claim that truth is a *practical* norm is much doubtful.

There is, therefore, no good reason to accept the view that truth is the intentional goal of belief in any teleological sense. It does not follow that the normative account is wrong. We have seen that it is usually associated to the idea that, in order to be responsive to the norm of truth, believers must recognise this norm and reflect upon them. But in order to believe that P, it is no more necessary to be aware that truth is the aim of belief than it is necessary to believe the proposition that *it is true that P*. Believing that P entails believing that P is true, but it does not involve any attitude towards the proposition ' 'P' is true'. It only involves an attitude towards P. Similarly, for believing that P, it is not necessary to believe that truth is what one is aiming at. Being able, at least tacitly, to recognise the epistemic norms of beliefs is certainly a necessary condition for having the concept of belief. But in order to be a believer, one does not need to attend reflectively to these norms. So the fact that belief aims at truth (in the normative sense) is one thing, and the fact that inquiry has truth as its goal is another thing. We could in this sense distinguish truth as the *distal* aim of belief and truth as the *proximal* aim. They are, of course, related, and one may think that if truth were not the norm for individual beliefs it could not be the general goal of inquiry. But the believer in an individual belief is not someone who contemplates the general goal of aiming at truth and acts upon it. Truth is her *proximal* aim.

5   Aiming at Truth and Aiming at Knowledge

The intentional or teleological interpretation of truth directedness is encouraged by a common picture of the nature of knowledge and justification. The picture is well articulated, for instance, by Laurence Bonjour:

> What makes us cognitive beings at all is our capacity for belief, and the goal of our distinctively cognitive endeavours is *truth*. We want our beliefs to correctly and accurately depict the world... The basic role of justification is that of a *means* to truth, a more directly attainable mediating link between our subjective starting point and our objective goal.... If epistemic justification were not conductive to truth in this way, if finding epistemically justified belief did not substantially increase the likelihood of finding true ones, epistemic justification would be irrelevant to our main cognitive goal and of dubious

---

[10]See Owens[Owens, 2000, pp. 108–109].

worth. It is only if we have some reason to think that epistemic justification constitutes a path to truth that we as cognitive human beings have any motive for preferring epistemically justified beliefs to epistemically unjustified ones. Epistemic justification is therefore in the final analysis only an instrumental value, not an intrinsic one. [Bonjour, 1985, pp.7–8]

This familiar picture is encouraged by two related ideas. One is the traditional definition of knowledge as justified true belief. The second is precisely the teleological conception of truth as the 'end' or 'ultimate goal' of inquiry. Given that truth is that ultimate goal, justification operates as the means towards that end. As we have seen, it is easy to transpose this picture, which holds for inquiry, hence for the formation of beliefs in general, to the attitude that a subject has when he forms a particular individual belief.

The picture has been most criticized, but for other reasons than those which interest me here. It is criticised by those who consider that the gap between justification and truth cannot be so large that truth could be considered as a goal. If one subscribes to an anti-realist theory of truth in particular, there should not be much difference between truth and justification. In this spirit for instance, Richard Rorty has attacked the claim that truth could be the end of inquiry. How, he asks, could one aim at something which one, by definition does not know? An aim, by definition is something which one can at least figure out. And by definition the ultimate end of inquiry is unfathomable [Rorty, 1995], see also [Heal, 1987]. But whether or not we agree that truth is a goal in the sense of an intrinsic value, these debates are besides the point. For they all presuppose that the relevant sense of 'aim' in the phrase 'beliefs aim at truth' is the intentional sense. But I have argued that this is wrong. The relevant sense is the normative sense. But the normative sense does not require at all that truth be conceived as a goal or as a value. It just requires that the norm of truth plays a constitutive role in the formation of beliefs, without the subject needing to aim, in the teleological sense, at a certain objective. It is an *a priori*, conceptual, requirement. To reinforce this point, we need to try and specify more what the point of belief is. And the point is not simply to believe only truths, but to be able to *know* them.

I have rejected the intentional account of aiming at truth by arguing that it wrongly equates believing to purposive attitudes such as guessing. But I have not explained why there should be a difference in the respective norms which govern the two kinds of states. For we have agreed that truth governs guessing as well as it governs belief. So what is the difference?

The difference lies in the fact that belief is in fact governed by another norm than the norm of truth (NT), although a quite related one. The norm in question is the norm of knowledge. When we believe that P, we aim at knowledge as much as to truth. For we do not need our beliefs simply to be true. We need them also to be—in some sense-justified or reliable. Suppose that I come to believe that P on the basis of my guessing that P. Then even if my guess turns out to be successful—that if it is indeed the case that P—my belief is true, but does it amount to knowledge? Intuitively no. The reason why is familiar from Gettier's problem and by the definition of knowledge as the exclusion of relevant alternatives: guessing that P is compatible with too many alternatives which could have made it the case that one does not know that P. Recently several writers have suggested that knowledge, not truth, is the standard of correctness for belief. A belief is botched, even if it happens to be true, unless it amounts to knowledge. In this sense belief aims at knowledge ([Williamson, 2000, p. 47, p. 208]; [Peacocke, 1999, p. 34]).

This suggests the following reformulation of our basic norm for belief as a norm of knowledge:

(NK) For any P, believe that p only if you can treat p as if you knew p

(NK) is distinct from (NT), but it is stronger, and entails it. If knowledge sets the standard for belief, given that knowledge entails truth, truth is still—in the normative sense—the aim of belief.

Why should (NK) rather than (NT) give a better account of the norm of truth for belief? The answer is simple. Belief aims at truth, in the sense spelled out by (NT), but, by definition, beliefs can be false: what I believe, from my own perspective to be true might not be true. This is unlike knowledge, which by definition implies the truth of the belief. In Williamson's terms, knowledge is a factive attitude [Williamson, 2000, pp. 33–45], like seeing that P or perceiving that P (to perceive that P entails P, to see that P entails P). Knowledge is the most general among factive attitudes. Now by definition 'believing truly' does not entail 'knowing' and is not factive [Williamson, 2000, p. 39]. But if our beliefs were true, in the sense that it would be sufficient for them to produce the truth of their propositional objects, they would be knowledge. It is in this sense that knowledge sets the standard for belief. It is because knowledge, and other factive attitudes, imply this necessary relationship to truth that they can serve as a model for belief, which is thus imperfect knowledge.

If one adopts this view, many things fall into place. For instance it is easy to understand that belief is not voluntary if we see that beliefs aim at knowledge, for it is in general impossible to know something at will [Williamson, 2000, p. 46]. On this view too, failures to comply by the norm of truth are also failures to know. If, in a given circumstance, I feel that I ought not to believe that P because P is false, it is also a case where I realise that I cannot know that P. The norm of knowledge (NK) is in this sense more fundamental than the norm of truth (NT). [11]

---

[11] I therefore agree with Wedgwood [2002] that the normative sense (NT) is the correct specification of the aim of belief. But I disagree with him that this norm is more basic than the norm of knowledge

I cannot here argue directly for the view that knowledge is the aim of belief, but the claim can be reinforced by answering two *prima facie* objections to it. Both threaten to render the proposal empty.

The first objection can be formulated thus. Suppose that knowledge, instead of truth, sets the standard for correct belief. Suppose too that knowledge is defined in the traditional way as justified true belief. But then we cannot *explain* the standard or the epistemic norms for belief, for the traditional definition is precisely supposed to give us a definition of knowledge in terms of justification. In general it is precisely for this reason that most epistemologists consider that truth is the external goal of the enterprise of knowledge. For if knowledge, instead of truth, were the goal, we could not analyse knowledge in terms of justification, for it would involve the very concept to be analysed in the *analysans*. [12]

The second objection is that if one takes knowledge, defined as justified true belief, as the aim of inquiry, then there cannot be any difference between knowledge and true belief. For justification can only be a means towards truth as the end of inquiry. Hence the *telos* of inquiry must be true belief, not justification. So if we take knowledge to be the *telos* of inquiry, there should be no difference between knowledge and true belief. But this is absurd, for if it were correct there would be no distinction between knowledge and successful guessing. [13]

The answer to these objections is that if we want the epistemic aim of belief to be knowledge, we have better *not* define knowledge in terms of justified true belief. And indeed this is precisely the line taken by Williamson. If knowledge

---

[Wedgewood, 2002, pp. 289–90].

[12] The objection is well formulated by David [2001]:

> Although knowledge is certainly no less desirable than true belief, the knowledge-goal is at a disadvantage here because it does not fit into this picture in any helpful manner. Invoking the knowledge-goal would insert the concept of knowledge right into the specification of the goal, which would then no longer provide an independent anchor for understanding epistemic concepts. In particular, any attempt to understand justification relative to the knowledge-goal would invert the explanatory direction and would make the whole approach circular and entirely unilluminating. After all, knowledge was supposed to be explained in terms of justification and not the other way round. This does not mean that it is wrong in general to talk of knowledge as a goal, nor does it mean that epistemologists do not desire to have knowledge. However, it does mean that it is bad epistemology to invoke the knowledge-goal as part of the theory of knowledge because it is quite useless for theoretical purposes: The knowledge-goal has no theoretical role to play *within* the theory of knowledge. [David, 2001, p. 154]

[13] Actually some writers, most notably Crispin Sartwell [1992], do not take it to be absurd; on the contrary they claim that knowledge is mere true belief on the basis is this argument. For a discussion of Sartwell's argument, see Olsson [2003]. Sartwell's argument is the product of *both* maintaining that truth is the aim of inquiry and that knowledge is the aim: given that they have the same aims, it follows that they are the same thing! One philosopher's *modus tollens* is another philosopher's *modus ponens*. Olsson shows that there is a *reductio* of the proposed thesis that knowledge is merely true belief, but that this reductio is not so easy to formulate.

is the aim of belief, knowledge must not be defined as the product of two factors, justified belief and truth; it must be taken as a primitive state, in terms of which belief is to be understood, and not the other way round. To believe that P, on such a view is to have an attitude towards P which one cannot discriminate from knowing, but which might fall short of being a state of knowing. To believe that P is to treat P as if one knew P, for all one knows [Williamson, 2000, p. 40].

6 Conclusion

I have examined three senses of the ambiguous phrase 'aiming at truth'. The causal one is insufficient to characterise belief. The normative sense is the correct one, but I have argued that it should not be understood in any intentional or teleological sense. I have also argued that the norm of truth for beliefs should better be taken to be the norm of knowledge (although the latter entails the former). In taking knowledge as the aim of belief, we must not be understand (NK) as specifying an *external* goal of belief, in a teleological sense. This picture is precisely the one which I have rejected by criticising the intentional interpretation of aiming at truth. The aim of belief is not external to the attitude of believing, but internal to it. This means that the question of the nature of belief, and of the norms which govern it, is independent of the question of the value of truth as the goal of inquiry.

Appendix

I have not examined here what seems, *prima facie*, to be an important objection to the very idea that truth could be the aim of belief, even in the normative sense here adopted. It is the objection that beliefs could only aim at truth if they were taken to be *categorical* or *full* beliefs. But if, in addition to full beliefs, which are by definition evaluated as true or false, there were also *partial* beliefs or *degrees of belief*, then the claim that truth (knowledge) is the aim of belief would be in jeopardy. This appendix tries to address in outline such an objection.

A Bayesian, who holds that there are no full beliefs, but only partial beliefs (or that full beliefs are only limit cases of partial ones) cannot agree that the norm of truth (NT), or even less (NK) is the norm of rational belief. He will take these norms as typical of dogmatic epistemology . In Ramsey's [1926] words, he will take the logic of 'coherence' to be more important than 'the logic of truth'. Moreover, he will deny that in believing we are responsive solely to a *theoretical* norm of truth or of knowledge. Belief, for the Bayesian, in so far as it obeys the laws of probability, is as much responsive to the norms of *practical* or *prudential* rationality as it responsive to the norms of theoretical rationality. The degree of a belief is defined by its betting quotient, and in so far as the agents expects to maximise his expected utility, if his partial beliefs violate the laws of probability, and are in this sense incoherent, a Dutch book can be made against him. The Bayesian analysis is a version of what I have called above the causal account of truth-directedness, except that it precisely gets rid of any notion of aiming at truth or of being directed towards truth. On the pragmatic analysis of belief championed by Ramsey and Jeffrey, ' the kind of measurement of belief with which probability is concerned is a measurement of belief qua basis of action' ([Ramsey, 1926, 1990, p. 67]; [Jeffrey, 1992]).

Now if belief is not an all-or-nothing affair, and if beliefs obey the norms of practical rationality, then it seems that the Bayesian analysis is not only incompatible with the one proposed here, but also that the latter is just wrong. But it is not obviously so. James Joyce [1998] addresses directly this problem, and proposes what he calls a 'non pragmatist vindication of probabilism', which he claims to be compatible with the view that beliefs are partial.

Joyce starts by a reformulation of the norm of truth as

> The Norm of Truth (NT): An epistemically rational agent must strive to hold a system of full beliefs that strikes the best attainable overall balance between the epistemic goal of fully believing truths and the epistemic evil of fully believing falsehoods (where fully believing a truth is better than having no opinion about it, and having no opinion about a falsehood is better than fully believing it).

This is an explicitly teleological formulation, but this need not detain us for the purposes of this discussion. Now is NT really incompatible with the idea that what we should aim at is not truth or full true beliefs, but only high degrees of credence or degree of subjective probability? No, for it could be said that in aiming at high degrees of credence is in fact aiming at truth. Likewise the lower degree of credence we give to a falsehood, the better we have achieve the goal of avoiding false belief.

So Joyce reformulates the Norm of Truth as

> The Norm of Graduational Accuracy (NGA): An epistemically rational agent must evaluate partial beliefs on the basis of their graduational accuracy, and she must strive to hold a system of partial beliefs that, in her best judgement, is likely to have an overall level of graduational accuracy at least as high is that of any alternative system she might adopt.

But this is not enough. For we have to ensure that, when forming their degree of beliefs, agents only frame them in terms of the *epistemic value* of having a high degree of credence. But it is by no means guaranteed by the pragmatic Ramseyan scheme, since by definition degree of belief is as sensitive to high expected utility as it is sensitive to high degree of credence. We have to ensure that in some sense there is *coincidence* between beliefs that elicit a high degree of credence and beliefs which elicit a high expected utility. In other words , the aim of 'truth', or of the high credence which approximates truth in the best possible way has to be aligned on the aim of desirability, or it has to be pure, so to say, whereas in the classical Ramseyan picture, it is always mixed.

Of course a pragmatist about reasons for belief will consider that the Bayesian scheme is perfectly in order if it mixes high credence function and desirability function, for it is precisely the gist of a pragmatist conception of values that the value of truth is a utility value, and that in some cases utility can override pure epistemic worth. In order to avoid this result we have to ensure that the two kinds of values either coincide or can be separated sharply so that only the epistemic goal is satisfied.

In other words, we have to make a distinction, which the Bayesian Dutch book argument does not make, between prudential or pragmatic reasons for believing (believing that P when it pays to believe that P) and epistemic reasons for believing.

In order to show that we can aim for a purely epistemic goal while at the same time ensuring that our degrees of belief are measured by probability functions, we must restrict the class of functions in a certain way. And Joyce shows that there are such functions, the *Brier rules*.

Now, suppose that there as such non pragmatic rules of accuracy satisfying the norm NGA as an analogue of the norm of truth. The problem, raised by Gibbard [2004] about this scheme, is: does this rule capture the pure concern with the truth? It is not evident. The concern for truth is represented only by a small subset of the possible functions.

Joyce [1998] gives a way of narrowing down the possible functions so that pure concern with the truth is captures. Joyce proves that graduational accuracy can be measured by a function which respects the following conditions: *structure* (accuracy should be non negative, i.e. small changes in degree of belief should not engender large changes in accuracy), *extensionality* (there is a unique correspondence between the degrees of credence of a person and the truth values of the propositions she considers), *dominance* (the accuracy of a system of degrees of belief is an increasing function of the believer's degree of credence in any truth and a decreasing function of her degree of confidence in any falsehood), *normality* (differences among possible worlds that are no reflected in differences among truth values of propositions that the agent believes should have no effect on the way in which accuracy is measured), *weak convexity* (if a certain change in a person's degrees of belief does not improve accuracy then a more radical change in the same direction should not improve accuracy either), and *symmetry* (when two degrees are equally accurate, there is no ground, based on consideration of accuracy to prefer one to the other). I cannot deal with these conditions here. But when accuracy is measured is this way, Joyce shows that conformity with the axioms of probability is a norm of epistemic rationality, whatever its prudential merits or dismerits can be. In other words, this forces the Bayesian into the straightjacket of the disinterested research for truth, and vindicates probabilism *non pragmatically*.

Gibbard [2004] argues that it is not evident that we can isolate pure concern for truth. What distinguishes the Brier score rule is the equal urgency of getting the credences right at every probability from 0 to 1. Then the difference between a credence of 0.47 and a credence of 0.56 is as urgent and important as the difference between 0.90 and 0.99. But that sounds odd. As Gibbard notes, the accuracy in degrees of belief can matter in the case of a scientific hypothesis—such as the hypothesis of Continental Drift in geology—after it is has been widely established (some residual doubt can remain), but it does not matter in case of acting on one's beliefs (it does not matter whether I have 0.63 or 0.62 chances of being hit by a car when I cross the street; in either case the risk is high, and I'd better not cross the street). Gibbard objects that there is no reason, even in the case of the purest and the most disinterested scientific beliefs, to suppose that they will not affect our actions. Indeed on the classical pragmatic picture, the truth of our beliefs guarantees the fact that they guide our actions. How can we be sure that the pure aim of truth is our only concern? At best, it is our concern only when in some way—that determined by the special score functions or by Joyce's non pragmatic vindication—concern for truth coincides with utility. But that is not guaranteed to be always the case. In other terms, even when one measures beliefs in a way which would ensure that they respect the Norm of Graduational Accuracy, we have no guarantee that it will be always so. At most, as Gibbard says, pure concern for truth can have a *side value* for an inquiry which is not governed by a concern for truth as a *guidance value*. There is no 'epistemic purity'.

The moral, it seems, is that truth is not the only guide we have, it is not the only goal, and it being essentially tied to action, our beliefs do not manifest a pure concern for truth. But the question whether truth is an intrinsic value or not was not, remember, our question. The norm of truth (NT) as I have formulated it does not say that truth or knowledge are the only values. It only says that the norm of truth is constitutive of being a belief. So it seems to me that the discussion above is not affected by the answer that one gives to the question whether we should formulate beliefs as an all or nothing or as a graded affair.

Acknowledgements

Earlier versions of this paper have been read at the University of Konstanz in January 2003, at the University of Rome in May 2003, at the University of Nottingham in November 2003 and at Stockholm University in March 2004. The paper was finished when I was a fellow at the Centre for Advanced Studies, Norwegian Academy of Sciences, Oslo. I thank the Centre for its hospitality and support. I thank for their comments Arthur Merin, Wolfgang Spohn, Gereon Wolters, Erik

Olsson and Simone Gozzano. I thank Peter Rosner and Woldek Rabinowicz for their comments, and Paul Noordhof, Gregory Currie, Olav Gjelsvik and my colleagues in Oslo for discussion of this paper. I thank also Huw Price and Isaac Levi who communicated to me their relevant papers [Price, 2003; Levi, 2002] but which I could not really take into account here.

BIBLIOGRAPHY

[Anscombe, 1958] G. E. M. Anscombe. *Intention*, Oxford, Blackwell, 1958.
[Bennett, 1990] J. Bennett. Why is Belief involuntary? *Analysis* **50**(2), 97–107, 1990.
[Bonjour, 1985] L. Bonjour. *The Structure of Empirical Knowledge*, Harvard, Harvard University Press, 1985.
[Broome, 2000] J. Broome. Normative Requirements. In J. Dancy, *Normativity*, Blackwell, Oxford, 78–99, 2000.
[David, 2001] M. David. Truth as the Epistemic Goal. In Steup, 151–169, 2001.
[Engel, 1998] P. Engel. The Norms of the Mental. In L. Hahn (ed.) *The Philosophy of Donald Davidson*, La Salle, Ill, Open Court, 1998.
[Engel, 1999] P. Engel. Volitionism and voluntarism about Belief. In A. Meijers (ed.) *Belief, cognition and the Will*, Tilburg, Tilburg University Press, 1–17, 1999.
[Engel, 2000] P. Engel (ed.). *Believing and Accepting*, Kluwer, Dordrecht, 2000.
[Engel, 2001] P. Engel. Is Truth a norm? In P. Kotatko, P. Pagin and G. Segal *Interpreting Davidson*, CSLI, Stanford , 37–51, 2001.
[Engel, 2002] P. Engel. *Truth*, Acumen, Chesham, 2002.
[Gibbard, 2004] A. Gibbard. Rational credence and the value of truth, manuscript, 2004.
[Heal, 1987] J. Heal. The Disinterested Search For Truth. *Proceedings of the Aristotelian*, Society, LXXXVIII, 97–108, 1987–1988.
[Humberstone, 1992] L. Humberstone. Direction of fit. *Mind* **101**, 59–83, 1992.
[Jackson, 2000] F. Jackson. Non cognitivism, Normativity, Belief. In J. Dancy (ed.), *Normativity*, Blackwell, Oxford, 100–115, 2000.
[Jeffrey, 1992] R. Jeffrey. *Probability and the Art of Judgment*, Cambridge, Cambridge University Press, 1992.
[Joyce, 1998] J. Joyce. A non pragmatic vincidation of probabilism. *Philosophy of Science* **65**, 575–603, 1998.
[Levi, 2002] I. Levi. Seeking Truth. In Wolfram Hinzen and Hans Rott (eds.), *Belief and Meaning: Essays at the Interface*, Hänsel–Hohenhausen, 2002.
[Noordhof, 2001] P. Noordhof. Believe what you want. *Proceeedings of the Aristotelian Society*, 2001.
[Olsson, 2003] E. Olsson. Minimal knowledge and the *Telos* of inquiry, unpublished.
[Owens, 2000] D. Owens. *Reason without Freedom, the Problem of Epistemic Normativity*. Routledge, London, 2000.
[Owens, 2003] D. Owens. Does belief have an aim? *Philosophical Studies* **115**(3), 275–297, 2003.
[Peacocke, 1999] C. Peacocke. *Being Known*, Oxford, Oxford university Press, 1999.
[Price, 2003] H. Price. Truth as Convenient Fiction. *Journal of Philosophy* C, 4, 167–190, 2003.
[Ramsey, 1926] F.P. Ramsey. Truth and Probability. In *Philosophical Papers*, D.H. Mellor (ed.), Cambridge, Cambridge University Press, 1990.
[Rorty, 1995] R. Rorty. Truth as the Goal of Inquiry: Davidson vs Wright. *Philosophical Quaterly*, 1995.
[Sartwell, 1992] C. Sartwell. Why Knowledge is merely True Belief. *Journal of Philosophy*, LXXXIX, 4, 167–180, 1992.
[Searle, 1983] J. Searle. *Intentionality*. Cambridge, Cambridge University Press, 1983.
[Sobel and Copp, 2001] D. Sobel and D. Copp. Against direction of fit accounts of belief and desires. *Analysis*, 61.1., 44–53, 2001.
[Stalnaker, 1984] R. Stalnaker. *Inquiry*. MIT Press, Cambridge, MA, 1984.
[Smith, 1994] M. Smith. *The Moral Problem*. Oxford University Press, Oxford, 1994.

[Steup, 2001] M. Steup (ed.). *Knowledge, Truth and Duty*, Oxford, Oxford University Press, 2001.
[Velleman, 2000] D. Velleman. On the aim of Belief. In *The Possibility of Practical Reason*, Oxford, Oxford University Press, 2000.
[Walker, 1996] T. Walker. The voluntariness of judgement. *Inquiry* **41**, 3.97–119, 1996.
[Walker, 2001] T. Walker. Truth aimedness and judgement. *Ratio*, XIV, 1, 68–83, 2001.
[Wedgewood, 2002] R. Wedgewood. The aim of Belief. *Philosophical Perspectives* **16**, 267–297, 2002.
[Williams, 1970] B. Williams. Deciding to Believe. In *Progl blems of the Self*, Cambridge, Cambridge University Press, 1970.
[Williams, 2002] B. Williams. *Truth andTruthfullness*, Princeton, Princeton university Press, 2002.
[Williamson, 2000] T. Williamson. *Knowledge and its Limits*. Oxford Univeristy Press, Oxford, 2000.
[Winters, 1979] B. Winters. Believing at will. *The Journal of Philosophy*, **76**, 243–256, 1979.
[Zangwill, 1998] N. Zangwill. Norms and Mind: direction of fit and normative functionalism. *Philosophical Studies* **91**, 173–203, 1998.

Pascal Engel

UFR de philosophie
Université Paris IV Sorbonne
1, rue victor Cousin
75230 Paris cedex 05 France
and Center of Advanced Studies, Oslo

Email: pascal.engel@paris4.sorbonne.fr

Mechanisms of Truth-directedness Comments on Pascal Engel's 'Truth and the Aim of Belief'[1]

WLODEK RABINOWICZ

It is a rich and thought-provoking paper. In this short comment, I cannot hope to do justice to the many issues it raises. In fact, I haven't even had time yet to familiarize myself with much of the literature to which Pascal Engel refers. He does cover lots of recent work on the subject and his paper should be extremely useful for someone who wants to know what's happening in this field. In my comment, however, I will concentrate on just one issue: I will question Engel's scepticism concerning the possibility of interpreting the truth-directedness of belief in causal-functional terms.

Engel considers three types of interpretations of the idea that belief is truth-directed: the *causal-functional*, the *normative* and the *intentional* one. He argues that 'only the normative sense gives us the right account of the aim of belief'. The intentional interpretation is incorrect, and the causal-functional account, while correct, is 'insufficient to characterise belief properly' (p. 78).

Now, this last remark may be somewhat misleading. I think that what Engel is after in this paper is not so much a proper characterization of belief, but rather a proper characterization of the *aim* of belief, or—more precisely—that he is after an answer to the question *in what distinctive sense or senses, if any, belief can be said to be necessarily truth-directed*. The important constraint on such an answer would be that we should not rest satisfied with any 'non-distinctive' sense of truth-directedness—a sense in which belief would be just one type of propositional attitude among many that all are truth-directed. Thus, to take an example, a distinctive truth-directedness of belief cannot simply consist in the fact that *whenever we believe that p, we believe p to be true*. For the same would apply to lots of other propositional attitudes, including the conative ones: Whenever we desire that p we desire p to be true, and similarly for hopes, wishes, fears, and so on. In fact, Engel interprets the constraint of distinctiveness even stronger: An interesting sense in which belief is truth-directed should distinguish belief not just from

---

[1] See this volume, pp. 77–37.

Laws and Models in Science, 101–106.

conative propositional attitudes (such as hope or desire), but also from other types of *cognitive* propositional attitudes, such as imagining, supposing, wondering, and so on.

Now, can we provide a *causal-functional* account of such a distinctive sense of truth-directedness? As I understand Pascal Engel, he wants to deny this. I would, however, want to suggest that he may be too pessimistic on this point.

A causal-functional account of belief can be provided either in terms of the characteristic effects of that attitude or in terms of its characteristic causes. To put it differently, in such an account of belief, one can either concentrate on its constitutive causal role or on the causal mechanisms that lie behind our beliefs and make them suitable to perform the role they are supposed to play. In what follows, I will refer to these two kinds of accounts as the *output* account and the *input* account, respectively, even though I realize that these labels may be misleading. Needless to say, a complete causal-functional account would need to take into consideration both these aspects of belief, the output side and the input side, but in practice the two aspects often are considered relatively independently.

As concerns the output accounts, a typical approach of this kind takes beliefs to be our guides to action. Beliefs are the maps by which we steer, to use Frank Ramsey's famous metaphor.[1] As Engel presents this idea,

> an attitude is a belief only if it disposes a subject to behave in certain ways that would tend to realize his desires if the proposition towards which it is directed is true (p. 79).

As far as I can see, it is not the output-directed account that allows us to view the truth as the aim of belief. On the contrary, on such an account, it would be more natural to say that the aim of belief is *action*. To get hold of the causal-functional sense of the truth-directness of belief, one should use the input approach instead. The truth-directedness of belief is to be found in the causal mechanisms that regulate our beliefs and make them suitable for their causal role as guides to action. In this respect, I take my cue from David Velleman, who presents the following input perspective on belief:

> Belief aims at the truth ... in the sense of being constitutively regulated by mechanisms designed to ensure that it is true. Belief also bears a more fundamental relation to the truth, in that it is an attitude of regarding a proposition as true; but in this respect it is no different from other cognitive states, such as assuming and imagining ... What distinguishes a belief from other states that take their propositional objects as true is that, unlike assumption or fantasy, belief tends to

---

[1] See his 'General Propositions and Causality', Philosophical Papers, ed. By D. H. Mellor, CUP 1990, p. 146.

track what *is* true, when its regulatory mechanisms are functioning as designed. ... Beliefs thus aim at truth in the same sense that the circulation aims to supply body tissues with nutrients and oxygen. Not just any movement of fluids counts as the circulation, but only those movements which are under the control of the mechanism designed to direct them at supplying the tissues. Hence the aim of supplying the tissues is constitutive of circulation, just as the aim of being true is constitutive of belief.[2]

According to Velleman, it is truth-directedness in this causal-functional sense that lies at the basis of the *normative* truth-directedness of belief: If it is constitutive of belief that it is an attitude regulated by mechanisms designed to guarantee truth-tracking, if the mechanisms work properly, then the norm of truth tracking for that attitude can be traced back to its constitutive features: Believing falsehoods would be a sign that belief is not functioning properly. Thus, the norm of truth applies to belief in more or less the same way as the norm of pumping blood applies to the heart: A heart that does not pump blood is not functioning properly.

Engel (pp. 84f) classifies Velleman as one of the prime representatives of the *intentional* interpretation of the truth-directedness of belief and supports this by some quotations from Velleman's recent book. On the intentional interpretation, Engel says, 'believing is identified with (a) the conscious recognition of the basic norm of truth and (b) the intention to respect and maintain this norm in the formation of one's beliefs'. (p. 84) On that interpretation, then, the truth aim of belief is simply identified with the truth aim of the believer.

Now, Engel rightly objects to the intentional interpretation of truth directedness: To begin with, it is not obvious that beliefs can be subject to intentions, not directly at least. Unlike actions, beliefs are not under our voluntary control. But quite apart from that issue, it seems that there could be believers who have no intention whatsoever of getting at truth and of avoiding falsehoods. They might have no aims for their beliefs at all. Or what they might care about is just that their beliefs are advantageous to them in terms of increased well-being. In fact, some very primitive believers may not even have the notion of belief at their disposal. They need not be sufficiently self-reflexive to be aware of having beliefs.

However, I think it is incorrect to ascribe the intentional interpretation to Velleman. In fact, he is explicitly opposed to the intentionalist view. Here is a couple of quotes that bear out this claim:

> consider that case of belief. Here the ... role [of a regulatory mechanism] may be played by the subject's desire to arrive at the truth; but it may also be played by sub-personal cognitive systems that are designed to track the truth, independently of the subject's desires. Truth

---

[2]D. Velleman, *The Possibility of Reason*, OUP 2000, Introduction, p. 17.

must be the aim of belief, but it need not be an aim on the part of the believer; it may instead be an aim implicit in some parts of his cognitive architecture. (*ibid.* p. 19)

To say that our attitude toward a proposition is partly constituted by the aim or purpose with which we accept the proposition is not to say that the aim is itself an attitude of ours, or that acceptance is an action. This point cannot be overemphasized. Acceptance is a mental state whose aim may be emergent in the cognitive mechanisms by which that state is induced, sustained and revised. (p. 184)

A person can also aim cognitions at the truth without necessarily framing intentions about them. Suppose that one part of the person—call it a cognitive system—regulates some of his cognitions in ways designed to ensure that they are true, by forming, revising, and extinguishing beliefs in response to evidence and argument. Regulating these cognitions for truth may be a function for which the system was designed by natural selection, or by education and training, or by a combination of the two. In any case, the system carries out this function more or less automatically, without relying on the subject's intentions for initiative and guidance. (p. 253) [Engel refers to this last passage in his paper, but he still takes Velleman to be an intentionalist!]

To summarize, then, Velleman manages to provide a causal-functional account of truth-directedness by changing focus from the causal role of beliefs to the regulatory mechanisms which make beliefs suitable to play the role in question. This seems like a very plausible view to me and I wonder why Engel has not given such an input-oriented account more attention in his paper. To be sure, he does mention it in passing. At one point, at the very beginning of the section on causal-functional accounts, he points out that beliefs are 'the kind of mental states which register true information. Of course they often fail to perform this function, which it is nevertheless their normal function'. (p. 79) And immediately thereafter he connects this observation with the well-known idea of the 'direction of fit': Beliefs are attitudes that are supposed to fit the world and not vice versa. But for some reasons that I don't quite understand, he doesn't try to pursue this idea any further. It would be interesting to know why.

Let me speculate a little about possible drawbacks of Velleman's suggestion.

(i) It might be questioned whether the causal-functional sense of truth-directedness is distinctive enough. While neither conative attitudes nor such cognitive attitudes as imagining or supposing are truth-directed in Velleman's sense, what about an attitude such as guessing? Engel takes up the case of guessing in his paper in connection with the intentional interpretation: he argues that the intentionalist construal of truth-directedness would not allow us to distinguish between believing and guessing. On that interpretation, both presuppose the intention of getting at truth. Now, a similar point applies to the causal functional interpretation: both belief and guessing are regulated by mechanisms designed to increase the probability of truth-tracking. Consequently, this sense of truth-directedness does not seem to be distinctive enough.

I don't think, however, that this objection is especially worrying. To begin with, why should we require that belief must be the only kind of attitude that is truth-directed? Surely, guessing is another such attitude, but on the other hand — as we have seen — there are lots of attitudes that are not truth-directed in the relevant sense. Secondly, in a well-functioning agent, guessing is regulated by mechanisms that are less exacting than those that regulate belief: For guessing, the expected degree of truth-tracking is lower. Which means that, in this causal-functional sense, one can distinguish the more exacting kind of truth-directedness that characterizes belief from the less exacting one that characterizes guessing.

(ii) One might question whether truth-directedness in Velleman's sense is a *necessary* feature of belief. Velleman certainly thinks so: he takes belief to be *constitutively* regulated by the mechanisms that are designed for truth-tracking. However, as we have seen, from the causal-functional perspective, belief can in principle be given a pure *output* account, in terms of its action-guiding function. This makes it possible, one would think, to envisage a radically non-rational believer who is guided by beliefs in his actions, but whose beliefs are not in any way regulated by truth-tracking mechanisms. Needless to say, this possibility of a radically non-rational believer is deeply unrealistic, since believers of this kind would not survive for very long in this world. But still, such a possibility, if coherent, would mean that truth-directedness is not an necessary feature of belief.

Now, someone like Velleman could say, I suppose, that what we have here is not just an unrealistic possibility. Instead, it is a fundamental *im*possibility, excluded by our concept of belief. Thus, on this view, the attitudes that would guide such a radically non-rational agent, who is responsive neither to evidence nor to arguments and who lacks any intention to arrive at truth, would *a fortiori* not be genuine beliefs. However, to insist on the impossibility of radically non-rational beliefs appears to me rather dogmatic. In order to talk about the truth as the aim of belief, it is quite sufficient, I think, to regard truth-directedness as a feature that beliefs must satisfy in real life, whether or not that feature is a part of our concept of belief.

Wlodek Rabinowicz
Department of Philosophy
Lund University
Kungshuset, Lundagøard
222 22 Lund, Sweden
Email: wlodek.rabinowicz@fil.lu.se

Beliefs as Mental States and as Actions: A Comment on Pascal Engel

PETER ROSNER

Pascal Engel characterises holding a belief as an unintended mental state bearing a close relation to knowledge.[1] He thus gives the rather metaphorical phrase *beliefs aim at truth* a precise meaning. Having a belief has normative implications, namely how different beliefs interact and how a person holding a belief should react in face of evidence relating to the belief. He calls that *the norm of truth*. Basically, beliefs should be treated as propositions in science. Engel mentions the *modus ponens*, empirical evidence contradicting a belief. Furthermore, he distinguishes strictly between beliefs and assertions, guesses, judgements and other mental and cognitive activities which, superficially seen, seem similar to having a belief. They are all intentional. However, if beliefs were intentional one had to know beforehand why one wants to believe P, which in turn presupposes knowledge. Thus one cannot believe at will. There are reasons to believe P and that has normative implications, but without reasons one cannot make oneself believe P.

Clearly, not all utterances which begin with 'I believe that. . . ' are covered by Engel's analysis. For example, if somebody says, *I believe that peace would prevail if only all people of good will were to work together* that would hardly qualify as a belief in Engel's sense. A person who makes such a statement may consider it true and may act on the basis of this belief. But we cannot say that the content of this phrase expresses knowledge in any sense. In this context to believe something means rather having faith in something. However, that does not create a problem for the analysis of beliefs in propositions typical for scientific inquiries as long as one is conscious of the limitations therein.

I.

Is it really possible to strictly distinguish between believing as a state of mind and actions like judging, accepting, inquiring? If believing P is a mental state with normative consequences—for example to revise it in case of evidence coming up contradicting P—we have to assume that believing P is the result of some intentional activity—for example having revised an earlier belief O due to some

---

[1] Pascal Engel *Truth and the Aim of Belief*, this volume.

evidence having contradicted O. Having the belief P is a moment in a sequence of having beliefs $\ldots N, O, P, Q, R, \ldots$ precisely because of the normative implications of a belief which are elaborated in Engel's contribution. Though it is correct that one has to have the belief P if one believed in O and was confronted with certain data and we can therefore say a person is forced to believe P, that belief cannot be conceptualised by leaving out intentions, namely to be prepared to look out for evidence which may contradict O or to think about other beliefs contradicting O, the normative dimension of beliefs. The distinction between having a belief as an unintended state of mind and as result of having formed a belief is not as strict as Engel considers it to be.

That can be seen by considering the imperative NT*, namely *for all P, if P is true, than you ought to believe that P*. Engel rejects that with the argument that it would imply to believe unimportant information. For example, if a weird gardener tells me that he has counted the blades of grass on a lawn and that there are precisely 235451 of them, I should not believe it. But what does it mean not to believe it? I surely should not waste my time to reflect whether it is true or take the pain of counting them unless there is a special reason to obtain information about the number of blades of grass on a lawn. However, that would imply that there is a cause for forming the belief.

The point is that there are different meanings for the statement, I do not believe P, namely I believe non-P and I have no beliefs concerning P. Consider the following:

I do not believe that the Dollar will rise above the Euro within a year.

That can mean:

(i)   *I believe that the Dollar will remain below the Euro for at least one year.*

But it could also mean:

(ii)   *I have no beliefs concerning the Dollar/Euro parity for the future.*

Whereas for (i) I should have reasons for holding this belief, for the latter it is not necessarily the case, as it can mean:

(iia) *The theory of exchange rates I consider true and all the information I have make me believe that it is impossible for me to form any belief about the movement of the exchange rate.*

For example: The exchange rate is manipulated by the big banks, by the angels in heaven, by the Jews, the American president, al-Quaida etc., and I have no information about their intentions.
But it can also mean:

(iib)   *I do not bother to form a belief about the Dollar/Euro parity.*

Stating (iia) is having a belief and Engel's arguments are valid. But this is not the case for (iib). I have no beliefs concerning the development of the Dollar/Euro parity, because it did not occur to me to give a thought on the development of it. I may consider the Dollar/Euro parity as unimportant as the number of blades of grass on a lawn. Therefore even if one accepts that beliefs are not under voluntary control, namely believing P or non-P, it is under control, at least to some extent, whether one forms a belief or not. Beliefs are not forced upon us. Having a belief is different from being falling ill from flu.

Note that there is nearly always a stream of information reaching us and we therefore constantly have to make decisions as to whether to give attention to them. A person who receives her income from speculating in the foreign exchange market will be active in revising her beliefs about the future Dollar/Euro rate, as well as an economist who attempts to make a relevant theoretical contribution to the theory of exchange rates. They only could utter (i) or (iia) when asked how the Dollar/Euro rate will develop during the next year, but not (iib). Both would claim that the belief about the one-year forward Dollar/Euro rate is forced upon them as they have a theoretical model of the exchange rate as well as some information to feed it. However, there are others who consider the future Dollar/Euro parity as informative as the telephone number of people listed in a directory they have no intention to call, or the number of blades of grass on a lawn.

That believing is not merely a passive indulgence is also important for another issue of Engel's contribution. He mentions that beliefs are closed under *modus ponens*. However, this can only mean that, if I believe P and I am confronted with P implies Q, I ought to believe Q or revise my belief concerning P. But do I have to believe Q if I am unaware of Q and/or I am unaware that Q is a consequence of P? If this were the case, I would have to believe all that is true given P, namely all possible consequences of P. If one believes a scientific proposition S, that is a proposition which is assessed by scientific methods, one clearly does not believe all what can be a consequence of S. Namely, one does not have to have formed a belief about a specific proposition T which might be a consequence of S. Actually it is impossible to have formed a belief about all T which are consequences of S, a kind of 'logical omniscience'.[2] If that were the case, it would not be possible to learn new things or to do any research.

I am not mixing up beliefs and assents or acceptances which Engel is anxious to keep apart. Indeed, what I presuppose is that to believe P implies that I am able to assent to P and to accept P with reasons only relating to my belief of P. The case he mentions, namely to accept a student in order to encourage her

---

[2] Pascal Engel, *The Varieties of Belief and Acceptance*, in: P. Engel (ed.), 'Believing and Accepting', Dordrecht: Kluwer, 2000, p. 20.

though one believes that she will not be a good student, my acceptance takes into account other things which are not related to the belief of whether the student will be a good student. If I accept a person as a student only in case I believe her to be a promising scholar, the act of accepting a student is a sign that I believe her to be a promising scholar. However, I may accept someone as a student for other reasons—to encourage her, for reasons of social justice, due to favouritism, etc. The case of a judge's verdict or that of a jury is similar. The verdict of non-guilty does not imply that the judge respectively the jury believed the defendant to be not guilty because the law demands that guilt must be proven beyond reasonable doubt.

II

As mentioned, Engel takes only beliefs as simple propositions, or to use Bernard Williams' words *empirical beliefs*, into consideration. But how should beliefs such as the following be treated?

(iii)   *I believe that detailed regulations of the economy have at best no results. It is better to leave markets alone.*

Believing that can be seen as having faith in (iii) rather than as a belief aiming at truth, the object of inquiry in Engel's contribution. Indeed, it would not be sensible to attach to this belief a numerical value between zero and one as probability that (iii) is true, such as it is sensible to assign to the belief in (i) or (iia) a subjective probability. I should revise these probabilities if new information arrives, for example, I read a convincing paper telling me that the theory I considered true has to be modified; or the European Central Bank takes an action I did not foresee, because when forming the probability for (i) I had made assumptions about likely actions of the European Central Bank. Revising the probability that (iii) is true after having read the latest issue of the *American Economic Review* is ridiculous although it surely contains a few articles contributing to the discussion about the effects of regulating markets. But can the belief expressed by (iii) be dismissed for the discussion about beliefs aiming at truth such as the belief concerning good-willing people and peace mentioned above?

I consider this question to be interesting.[3] For more than 200 years there were economists who assented to (iii) and others who contradicted it. I think it is justified to say that there were many who believed (iii) to be true and others who believed (iii) to be false. The question into the truth-value of (iii) was and still is paramount to economic analysis.

Of course, it might be the case that (iii) is not merely a proposition about the working of a market economy. If a person utters (iii) he may believe that government activities besides having beneficial effects on aggregate welfare can endanger

---

[3]My background is economics and one of my fields is history of economics.

freedom (as for example F. A. Hayek), or that the state is likely to be captured by special interests if it makes detailed regulations (as for example James Buchanan). However, if (iii) slightly rephrased, one can take account of that.

(iii′) *I believe that, all things considered, detailed regulations of the economy have at best no results. It is better to leave markets alone.*

Be that as it may, the belief stated in (iii′) aims at truth as well, or to use Engel's words aims at knowledge: economists want to know as to whether it is better not to interfere into the working of a market economy. They can support (iii′) or reject it with arguments. Indeed, there are economists who in their best research, theoretical and empirical, make the market economy look quite good. There are other ones who argue the contrary to be true in their research. One finds in both groups Nobel laureates, for example R. Lucas belonging to the first group and J. Stiglitz to the second. Actually one hardly can find economists who do not have a belief concerning (iii′).

The point is that most of them use the same theoretical instruments, as well as using the same data and the same methods to analyse the data. An economist accepting (iii′) either accepts the result of the scientific publications of the others who believe non-(iii′) as being correctly done, or, if not, are able to point to the errors and shortcomings made in the analysis. They have the same theoretical and empirical knowledge. However, they fundamentally disagree on the working of a market economy, though economists of both groups would claim that the theoretical apparatus they use—which is the same for both—is appropriate for the analysis of the working of a market economy. Therefore we cannot say that economists know that (iii′) is true respectively false, such as one can say that economists know that in competitive equilibrium prices equal marginal costs and that the compensated demand functions are negatively sloped; or that physicists know the Ptolemaic model to be wrong. (There may be people who, after having studied physics or astronomy, still consider the Ptolemaic model to be true, but they are considered being cranks).

It would not do to classify (iii′) as a proposition stating what one holds to be true, perhaps as a working hypothesis. It is also not merely an assumption, like the assumption of perfect competition or that there are no external effects of private actions which causes the decentralised allocation to be suboptimal. Such assumptions are specific to a model and do not express any belief about the world to be analysed. The belief expressed by (iii′) is rather related to the belief that the world can be adequately described by a model of perfect competition, or that by assuming that there are no external effects one does not leave out essential properties of the world as it is.

I suggest calling a belief like that expressed in (iii′) a *fundamental belief*. If somebody utters (iii′) she presents a belief of a different character compared when stating (i) or (iia). Unlike (i) and (iia) the belief uttered in (iii') is something like a *Weltanschauung*. Whether it is true is one of the big questions which provoke embittered discussions, often with deep analytic content only specialists are able to grasp but never can be finally answered. The beliefs (i) and (iia) are simple propositions which somebody can make after having given at least some thoughts about it. A research programme to evaluate these beliefs can be envisioned, which is not the case for the belief (iii').

But believing (iii') is not a simple feeling. When a person is asked why she believes (iii'), she will give reasons. These reasons are often beliefs as well. For example, if one states, that the high unemployment in some countries is evidence against (iii'), somebody believing (iii') may answer:

(iv)   *I believe that the protection of ongoing employment contracts in the law is a hindrance to the reduction of unemployment.*

Somebody believing (iii') and thinking about the labour market may form the belief (iv). An economist believing (iii') and specialising on labour markets may work many years in order to make the belief expressed in (iv) more precise, for example:

(v)   *There is evidence that by reducing severance payments by 50% unemployment can be reduced by 20%.*

Note that a belief like (v) is usually presented as the outcome of a research process. Actually the research process is the argument in favour of (v). Note further that (v) can be used as an argument in favour of (iii'), as it states that state interference in labour markets increases unemployment. However, economists rejecting (iii') hardly would start an inquiry into the working of labour markets with the result of (v).[4] The direction of the research of economists is formed by the belief in the truth-value of (iii'). J. Stiglitz would not develop models which provide arguments that a pure market economy can result in economic disaster if he believed in (iii'), nor would R. Lucas proceed with his research if he did not believe in it.

What would an economist believing in (iii') say if presented with an argument, that is a proposition S, against (iii') she accepts to be correct? In general she will not say, I change my belief, rather she would say: although S is true there is another proposition T showing that S is not generally valid or does not have the consequences as assumed. It is an astonishing fact about the development of economics that economists are able to agree on all technical aspects of the analysis,

---

[4] Karl Marx presents in *Das Kapital* the laws of the pauperisation of the working class and the tendency of the rate of profit to fall as conclusions within the economic theory he had developed. But he was convinced that these two tendencies are true long before he had developed his economic theory, as they appear already in his early writings.

such that it is justified to speak of economics being a unified science as it is the case in physics; however they agree to disagree on fundamental propositions. That remains a realm of beliefs.

These beliefs always were and still are of great importance for the development of economics. In which way they are formed, I do not know. Probably it has something to do with social and economic developments and personal experiences. The long depression of the 1930s made economists look for arguments why aggregate demand failures may occur. The arguments they developed, the Keynesian revolution, became common knowledge amongst economists. However, the inflationary period starting in the late 1960s manifestly showed that the set of propositions developed in the Keynesian Revolution was not the end of the story. It triggered the formation of new propositions showing the shortcomings of Keynesianism—Monetarism. For many, this looked like a return to the beliefs of the pre-Keynesian era, and it is probably true that many protagonists of the development of monetarism always were convinced (iii) being true. Nevertheless the arguments that were developed cannot be characterised as stating a belief like (iii). That is, economist believing in (iii) being false had to argue their case anew, namely to challenge the propositions of the Monetarists by the same methods and with the same theoretical apparatus by which they have been substantiated before.

Peter Rosner

Department of Economics
University of Vienna
Hohenstaufengasse 9
1010 Vienna
Austria.

Email: peter.rosner@univie.ac.at

How a Type-type Identity Theorist Can Be a Non-Reductionist: An Answer from the Idealizational Conception of Science

KATARZYNA PAPRZYCKA

Two constraints are at work in our philosophical conceptions of the mental. First, we think that the mental is somehow radically different from the physical. However, the cost of underwriting this intuition ontologically is the mind-body problem, whose solution is monism. This has in part motivated the second intuition, viz. that the world is ontologically homogenous—it is made of just one kind of stuff, matter. The most natural form of materialism is the type-type identity theory, according to which there is an identity not only between mental and physical event-tokens but also between mental and physical event-types. It has been widely agreed that this version of monism straightforwardly leads to reductionism, the conviction that higher-level sciences (in particular psychology) will be in the end reduced to physics. This in turn leads to a straightforward denial of the first intuition, that the mental is different from the physical.

One way of trying to reconcile both intuitions has been the position of non-reductive materialism. One of the best known proposals is Davidson's anomalous monism, one of whose distinctive claims is that there are no type-type but only token-token identities among the mental and physical events.[1] The reception of the token-token identity theory has been mixed. On the one hand, it provides a solid basis for the position of anti-reductionism and thus for reconciling both intuitions. On the other hand, however, it has been faced with objections such as the objection of type-epiphenomenalism, according to which if one gives up type-type identities, it becomes unclear how the mental as mental can be viewed

---

[1] I will not be concerned here with Davidson's second distinctive claim, which actually motivates his version of token-token identity theory. Davidson believes that there are no psychological (and psychophysical) laws in part because folk-psychological generalizations involve content-bearing propositional attitudes. Davidson provides no reasons, however, to believe that the science of psychology will be modelled on folk psychology. His paradigm case of psychological research is the study of the limitations of formal decision theory as a psychological theory, which seems rather special and not at all paradigmatic of the science of psychology as it stands, not to mention any future psychology.

Laws and Models in Science, 115–130.
© *2004, the author.*

as being causally efficacious at all [Soutland, 1976; Stoutland, 1985; Honderich, 1982; McLaughlin, 1989; Kim, 1993a].

It is clear that the issue at stake lies on the borderline between philosophy of mind, metaphysics and philosophy of science. Most of the current debate is carried out among philosophers of mind and metaphysicians. The aim of this paper is to show how the debate can be enriched by drawing on some developments in philosophy of science. I will show in particular that if one adopts the idealizational conception of science [Nowak, 1971; Nowak, 1977; Nowak, 1980], one will be able to understand, on the one hand, how one theory could be irreducible to another even though there are type-type identities among the theories' relevant predicates. Moreover, the proposal allows one to escape the charge of type-epiphenomenalism as well as Kim's challenge to non-reductive materialism.

1   A Methodological Interlude

There is a sense that the debate between reductionists (convinced that all science will be ultimately reduced to physics) and anti-reductionists (convinced that the thesis of reductionism is false) is deeply unsatisfying to anyone who has a deep respect for science and in particular for its ability to surprise us intellectually. This kind of philosophical legislation of what science will (or even may) do is perhaps particularly irritating with respect to psychology, which is by no accounts developed enough for us to even begin to see any overarching theories or even the direction for such, not to speak of the possibility of reducing them.

In this kind of situation, one faces a choice. One can try to find philosophical arguments for either of the sides—one can either become a reductionist or an antireductionist. Another option is to become an agnostic and refuse to take sides insisting on the need to look at the developments in the actual sciences before one could make the relevant judgments. This second option can take two flavors. One can become a passive agnostic and simply not get involved in the debate. But equally well, one can become an active agnostic and while refusing to take a stand one can try to contribute to the debate, in particular by multiplying the various theoretical possibilities.

This paper is written in the spirit of just such an active agnosticism. Moreover, such an attitude can help us to better understand the major change that took place on just this issue in the last century. While at the beginning of the 20th century, philosophical intuitions seemed to lie on the side of reductionism, the end of the 20th century has been taken over by a 'fashion' for anti-reductionism. The term 'fashion' is justified because the major shift in our positive attitude toward anti-reductionism has not been supported by proportionally good ways of understanding how anti-reductionism is possible (as the critics of anti-reductionism have been eager to point out, see e.g. [Kim, 1998]). This stands in stark contrast to the position of reductionism, which has been rather clearly understood (see e.g. Lewis

[1966; 1972]). However, if we accept the position of active agnosticism (as well as if we assume that the philosophical community has more or less consciously accepted such an attitude), then we can understand its predominant anti-reductionism as more than a fashion: it is precisely because reductionism is better understood while anti-reductionism is faced with numerous problems that we need to invest our energy into a better understanding of how anti-reductionism is at all possible.

2   E. Nagel's Ontological Reduction

We will begin with the widely accepted starting point for both reductionist and anti-reductionist conceptions, viz. E. Nagel's [1961] conception of reduction. Nagel distinguishes between homogeneous reduction (where the theories reduced use the same concepts, as he conjectured is the case in the reduction of Galileo's law of free fall to Newton's law of gravitation, for example) and heterogeneous reduction (where the theories use different conceptual schemes, as is the case in the reduction of 'qualitative thermodynamics' to molecular thermodynamics, for example). If our target is the reduction of a (developed) psychological theory to a (developed) physical or physiological theory, we should be focusing on heterogeneous reduction.[2]

E. Nagel lists numerous conditions both formal and empirical that are satisfied by the reduced and the reducing theories. They include the justification of the theories, their historical development, their systematization. One of the two central formal conditions is the requirement that the terms of the reduced theory that do not appear in the reducing theory must be expressed in terms of the reducing theory. Such relations between the terms of the respective theories are captured by the coordinating definitions or 'bridge laws'. According to the second main condition:

(C) the experimental laws of the reduced theory are logical consequences of the theoretical postulates and coordinating definitions of the reducing theory.

---

[2]I believe that the distinction has in fact only led to problems. Homogeneous reductions, which have been regarded by Nagel as so unproblematic, have in fact given rise to one of the most notorious problems in philosophy of science, the problem of incommensurability [Kuhn, 1962/1970]. However, there is a not frequently noticed though relatively easy way out of the problem once one only gives up on the distinction. It is relatively natural to think that the concepts of the reduced are being *explicated* in terms of the concepts of the reducing theories in the cases of heterogeneous reductions. If that thought is accepted, there is but a short step to thinking that the very same thing happens in the cases of so-called homogenous reductions. This, of course, raises the problem of what explication is (I make some suggestions on how to approach it in Paprzycka [1999] and [forthcoming]), but this is a worthwhile philosophical query in any case, and the benefit of this philosophical strategy is that it reduces one philosophical problem (incommensurability) to another—broader—one (the problem of the nature of explication). (This strategy has been partially endorsed in a rather not well-known paper by James Gaa [1975].)

It should be thus stressed that reduction so conceived is based on coordinating definitions, the so-called bridge laws, which express type-type relations (identity, or nomological equivalence) between concepts of the reduced and reducing theories.

3   Irreducibility I: Token-token Identity

Is it possible for there to be two true theories explaining the same phenomena, where neither one reduces to the other? Davidson [1970] offers a positive answer. Reduction will be impossible if the concepts of the theories in question cannot be related to one another by means of bridge laws, i.e. if there are no type-type identities, even though there are token-token identities among mental and physical events. Mental event-tokens just are physical event-tokens. But such event-tokens can be described by means of mental concepts or by means of physical concepts. These two conceptualizations cut the nature at different joints and cannot be identified one with the other. Psychological theories will be accordingly irreducible to physiological or physical theories.

It is thus clear that it is the rejection of type-type identity relations that underlies Davidson's anti-reductionism.

The Charge of Type-Epiphenomenalism. It has been objected that Davidson's view leads to a version of epiphenomenalism, viz. type-epiphenomenalism. [3] It is worth pointing out that the view is free of classical epiphenomenalism, according to which mental events are causally mute. According to the token-token identity theory, mental event-tokens are causally efficacious in exactly the way in which physical event-tokens are since mental event-tokens are identical with physical event-tokens. In this view, however, mental events are not causally efficacious qua mental or with respect to their mental properties.

Consider the following example: John shouts 'The world is wonderful' so loud that the glass in the window breaks, and his depressed neighbour interrupts her garden work to throw some wilted flowers through the newly opened spaces (i.e. the broken window). In this case, John's shout is the cause of at least two events— of the breaking of the glass, on the one hand, and of the fact that his neighbour becomes upset, on the other. In the first case, it is clear that only physical properties of the shout are in play—its rapidly rising amplitude. In the second case, other properties are in play—semantic properties (if the depressed neighbour disagrees with the content of the shout) as well as psychological properties (if she is appalled by John's state of mind thus manifested).

This example is just an illustration of the fact that causal relations hold not just between events (as is suggested by Davidson) but rather between events with

---

[3]F. Stoutland [1976] was the first to put forward this charge against Davidson's view. Later the objection has been presented in numerous versions by Honderich [1982], McLaughlin [1989], Kim [1993a].

respect to certain properties that the events exemplify. In the above case, the event of John's shouting is a cause of two different events with respect to two different properties. With respect to its rapidly rising amplitude, it causes the window glass to break; with respect to its content it upsets his already depressed neighbour.

While Davidson resists this move [1993], he has been—with good reason[4]— taken to privilege the physical properties and so understood as claiming that mental/physical events cause other mental/physical events *qua* physical, i.e. with respect to their physical properties. But if so, then there is no room in Davidson's theory for the claim that mental/physical events can cause mental/physical events with respect to mental properties since Davidson resists the identification of mental properties with physical properties.[5]

It is worth stressing here that the problem arises because Davidson gives up on type-type identities. He could claim that events could cause other events with respect to their mental properties if mental properties could be identified with physical properties. But this is precisely what he denies, and what underlies his anti-reductionism. We are thus faced with a dilemma: if we want to accept Davidson's anti-reductionism, we are committed to rejecting type-type identities, which leads to type-epiphenomenalism; to avoid type-epiphenomenalism, we would have to accept type-type identities but at the cost of rejecting anti-reductionism. It turns out that this is a false dilemma. We can accept type-type identities and find room for anti-reductionism.

4   Irreducibility II: Non-natural Type-type Identity

Before going on to suggest how one can find room for type-type identities and in principle irreducibility, it will pay to mention one other way of addressing the issue. Jerry Fodor [1974] has famously tried to resist the reduction based on the intuition made famous by Putnam [1967], viz. the multiple realizability of psychological concepts. In brief, Fodor argued that reduction requires that the bridge

---

[4]While Davidson resists accepting the 'x causes y with respect to property Z' idiom, he does accept the physicalistic Principle of the Nomological Character of Causality, according to which every causal relation between concrete events is subsumed under some strict law of physics. He thus opens himself up to the following argument. Take any causal relation between event-tokens, a causes b. Now, according to the Principle of the Nomological Character of Causality, there is a universal law which subsumes this causal relation, i.e. $(x)(y)(P(x) \rightarrow R(y))$, where $P(a)$ and $R(b)$. But now the critic will jump in and say that this is just what it means to say that 'a causes b with respect to property P'.

[5]There is some room open to manoeuver on Davidson's view. He could claim that the Principle of the Nomological Character of Causality does not settle it that there is *only one* law that subsumes a given cause (or even a given causal relation). This would make intelligible the distinction between a causing b in virtue of one property and a causing c in virtue of another property. But one would have to give up either Davidson's physicalism (the view that only physics is privy to universal causal laws) or his universalism (the view that causal relations are underwritten by universal causal laws) in order to be able to claim that an event can cause others in virtue of other than physical properties. (Moves in this direction have been suggested by McDowell [1985]).

laws register nomological relations between natural kinds and that any bridge laws between psychological concepts and physical concepts would not capture identities between natural kinds because the only way to think of psychological kinds in physical terms would be necessarily disjunctive and disjunctive kinds are not natural kinds.

Fodor's position has been challenged by Kim [1992] who argues that in the case where one 'higher' natural kind is a disjunctive kind based on some other 'lower' natural kinds, we would not have 'higher' laws involving the 'higher' natural kind, which would be irreducible to the 'lower' laws involving the 'lower' natural kinds (as Fodor thinks), but rather we would simply have to recognize that what we considered as the 'higher' natural kind is not a natural kind at all. This is because any causal power that this 'higher' natural kind has are entirely derivative from those of the 'lower' natural kinds. Kim recalls a case from the history of mineralogy to support this thesis. Jade was once thought to be a natural kind, but this view was rejected once it was discovered that jade is not a homogenous stone, but is made of two natural-kind stones, nephrite and jadeite, occurring in various combinations.

Fodor's [1997] response to Kim is to allow for the possibility of their being purely disjunctive kinds for which there are no empirical laws but Fodor argues that it would be simply question-begging to disallow the possibility of 'higher'-order natural kinds for which there are independent empirical laws. [6] The very same point can be, of course, levelled at Fodor from Kim's perspective. We arrive at a standstill.

Kim's Dilemma. The reasoning underlying Kim's [1989] could be, however, presented in the form a dilemma to the anti-reductionists. If it is indeed the case that there are psychological regularities that are irreducible to physical regularities, this means that the mental properties cited in such regularities have causal powers. The causal powers of those mental properties can be thought to be either dependent on the causal powers of the physical properties or to be independent of them. If the first option holds, if the causal powers of mental properties are dependent on the causal powers of the physical properties, we could not claim that any *sui generis* psychological regularities are discovered—psychology would then be reducible to physics if only locally (this corresponds to the jade case and such a position would be that of reductive materialism). If the second option holds, if the causal powers of psychological properties are independent of the causal powers of physical properties, this would be tantamount not to the position of non-reductive materialism but to the position of (property) dualism, which is well-known for its problems. In either case, it seems that the position of non-reductive materialism (also in Fodor's version, i.e. of non-natural type-type identity theory) is unstable—if one resists the reductive version of materialism, one seems to be committed to dualism.

---

[6]Similar distinctions have been in effect proposed by Block[1997].

5 Reduction in Terms of the Idealizational Conception of Ccience

Although much work has been done on homogeneous reduction in idealizational terms [Krajewski, 1977; Nowakowa, 1975; Nowakowa, 1994; Paprzycka, 1990], there has not been a systematic treatment of heterogeneous reduction.

For our purposes, we will simplify the discussion by simply assuming that the coordinating definitions express identities between the concepts of the reducing and the reduced theory. Some words of justification are in order. The assumption is not meant to be even suggestive of the way things are. It is likely to turn out to be false—there are a lot of intricacies here that we are simply pushing to one side. The justification for the assumption has to do with its place in the debate between reductionism and anti-reductionism: we are making the assumption that seems to be the safeguard of reductionism, and we will be showing that—under appropriate conditions—it can lead to anti-reductionism just as well.

Since we are applying the tools of the Idealizational Conception of Science, we will need to add another assumption to Nagel's assumptions, viz. that the theories in question are simple idealizational theories.[7] In addition (for details, see [Krajewski, 1977; Nowakowa, 1975; Nowakowa, 1994; Paprzycka, 1990]), we will assume that the theories t and T investigate the same magnitude and the space of factors considered by t to be essential to the investigated magnitude is a subset of the space of factors considered by T to be essential to the investigated magnitude. The cases where a reduction involves the addition of a new essential factor have been considered in the discussion of homogeneous reduction. Here we will only consider a case where the space of essential factors is identical between the two theories.

What certainly needs to be modified is condition (C). We will say that the idealizational theory t is reduced to the idealizational theory T in case:

(C$_i$) T together with appropriate coordinating definitions makes it possible to derive* the idealizational law of t as well as its concretizations.

Two points are in order. First, the term 'derive*' is meant to be noncommittal as to the exact nature of the derivation. It could be that sometimes the derived* theorems are logical consequences of the reducing theory, but they can also be approximations [Schaffner, 1967] or limiting cases thereof. We will simply not pursue this point further (and I will use 'derive' without the asterisk from now on) since this is quite a general problem for any account of reduction.

---

[7]The basic terminology of the Idealizational Conception of Science necessary for our purposes is presented in the next section where appropriate schematic examples are introduced. For a detailed presentation of the view, see Nowak [1977; 1980], for the summary of further developments, see Nowak [1992; 2000], and for various applications, see Nowak and Nowakowa [2000].

The second point, however, is crucial and it bears emphasizing. In the case of the idealizational construal of theories, there is a choice of how the derivational condition is to be construed. One could, for instance, demand that only the last concretization of the reduced theory t be derived from the reducing theory, but not that the idealizational law as well as all its concretizations be so derived. In such a case, however, it would be hard to speak of the reduction of the *theory*, one could at best speak only of the subsumption of the empirical results obtained by the theory to another. Part of what is involved in the reduction of one theory to another is not just that certain regularities that were accounted for by the reduced theory are also accounted for by the reducing theory, but also that some of the theoretical and explanatory work done by the reduced theory can be—to some extent—inherited by and preserved in the reducing theory. If the subsumption of empirical results were all that mattered then we might have to look for reduction relations between contemporary scientific medicine and 'witchcraft medicine'. On the positive side, some of the classical examples of reduction do conform to (C_i). The (idealizational) ideal gas law can be derived from statistical mechanics by means of appropriate coordinating definitions, and likewise can one derive its concretizations proposed by van der Waals (see Kuipers [1982; 1985; 1990]).

6   Irreducibility III: Essential Incompatibility

Intuitively: two idealizational theories can be proposed for the same domain of phenomena, which they can conceptualize in different ways. Even if the factors considered to be essential to the investigated magnitude may be identical, their essential structures will differ – what according to one theory is the most essential factor may be less essential according to the other.

Let us assume that we are considering two simple idealizational theories: theory t and theory T, whose domains are identical. The theory t investigates factor C, for which M, m_1 and m_2 are essential.[8] The *essential structure* of C is the hierarchy:

$$S_C : \quad \begin{aligned} & M \\ & M, m_1 \\ & M, m_1, m_2 \end{aligned}$$

where M is the principal (the most essential) factor, while m_1 and m_2 are secondary factors of diminishing essentiality.

---

[8]The notion of essentiality as well as that of the degrees of essentiality are primitive in the Idealizational Conception of Science (Nowak [1977; 1980]). Intuitively, to say that factor x is essential to factor y is to say that factor x influences factor y, and the degree of x's essentiality for y expresses the extent of such influence. However, these intuitive stipulations are far from satisfactory. Moreover, given the centrality of the concept, it is one of the main research projects to try to offer explications for it. With various degrees of success, some work in this direction has been carried out by Machowski [1990], Nowak [1989], Paprzycki and Paprzycka [1992], Paprzycka [forthcoming].

We will assume that t is composed of the following *idealizational law*

$$(t_0)(x)[m_1(x) = 0 \wedge m_2(x) = 0 \to C(x) = g_0(M(x))],$$

where the expression '$m_1(x) = 0$' is an *idealizing assumption*, whose effect is to assume that factor m_1 does not exert any influence on factor C, and the function g_0 expresses the relation between the principal factor M and the investigated factor C on the assumption that secondary factors do not exert any influence on C–g_0 is also said to express the *regularity*.

We will assume further that the idealizational law (t_0) has been concretized first with respect to the more essential secondary factor, viz. m_1, by postulating and testing the first concretization of the idealizational law:

$$(t_1)(x)[m_1(x) \neq 0 \wedge m_2(x) = 0 \to C(x) = h_{k1}(g_0(M(x)), k_1(m_1(x)))$$
$$= g_1(M(x), m_1(x))],$$

where the so-called *corrective function* k_1 shows how the thus far neglected factor m_1 influences the investigated magnitude, and the *directional function* h_{k1} shows how the corrective function modifies the regularity. When these two functions are superimposed, we can speak of the function g_1, which shows how factors M and m_1 affect the investigated magnitude.

We will also assume that the last concretization, which takes into account factor m_2 has been carried out:

$$(t_2)(x)[m_1(x) \neq 0 \wedge m_2(x) \neq 0 \to C(x) = h_{k2}(g_1(M(x), m_1(x)), k_2(m_2(x))$$
$$= g_2(M(x), m_1(x), m_2(x))],$$

where k_2 is the corrective function, h_{k2} is the directional function, and g_2 expresses the dependence of factor C on all factors essential to it. The last concretization of the idealizational law is also called the *factual statement* since it no longer applies to idealized models but rather to reality. The sequence of statements $(t_0), (t_1), (t_2)$ is called a *simple idealizational theory*.

We will assume further that theory T investigates factor D, for which factors N, n_1 and n_2 are essential. The theory T assumes the following essential structure S_D:

$$S_D : \quad N$$
$$N, n_1$$
$$N, n_1, n_2$$

We will assume that theory T is composed of the following idealizational law:

$$(T_0)(x)[n_1(x) = 0 \wedge n_2(x) = 0 \to D(x) = f_0(N(x))]$$

and its concretizations:

$$(T_1)(x)[n_1(x) \neq 0 \wedge n_2(x) = 0 \to D(x) = f_1(N(x), n_1(x))]$$
$$(T_2)(x)[n_1(x) \neq 0 \wedge n_2(x) \neq 0 \to D(x) = f_2(N(x), n_1(x), n_2(x))],$$

where f_0 expresses the regularity of theory T, while f_1 and f_2 express the dependence of the investigated magnitude on the appropriate essential factors.

Let us assume further that it was discovered that (a) the investigated magnitudes are identical

$$C = D,$$

and (b) that the remaining factors of theory t are likewise identifiable as factors of theory T. It is here that a crucial distinction arises. If the factors of both theories are identified in such a way that there is an isomorphism between the essential structures of the two theories, then the theories are *essentially compatible*; if there is no isomorphism between them, they are *essentially incompatible*.

Essentially Compatible Theories. In the above example, theories T and t will be essentially compatible if the following coordinating definitions are in order:

$$M = N$$
$$m_1 = n_1$$
$$m_2 = n_2.$$

In this way, the essential ordering of the factors is preserved. If so, then it is possible to derive (t_0) from (T_0), and the same holds for the respective concretizations. This structure is represented in Figure 1.

The case of essentially compatible theories exemplifies the relation between theories that corresponds to reductionist intuitions. If theories are essentially compatible, then it is possible to derive idealizational law of the reduced theory t from the idealizational law of the reducing theory T, and it is also possible to derive each concretization of the idealizational law of t from the respective concretization of the idealizational law of T. An example here would be the reduction of qualitative thermodynamics to statistical thermodynamics, where both the idealizational law (the ideal gas law) and its concretizations (van der Waals' corrections) can be derived from statistical thermodynamics (see Kuipers [1982; 1985; 1990]).

Essentially Incompatible Theories. In the case of essentially incompatible theories, the essential factors of theories t and T are related by means of coordinating definitions that do not preserve the isomorphism between the essential orderings of both theories (identifying principal factors of one theory with secondary factors of another). Let us take as a schematic example the following coordinating definitions:

$$M = n_1$$
$$m_1 = n_2$$
$$m_2 = N.$$

Given such coordinating definitions, it is impossible to reduce theory t to theory T (if we accept (C_i). Consider just the consequents of the idealizational laws first.

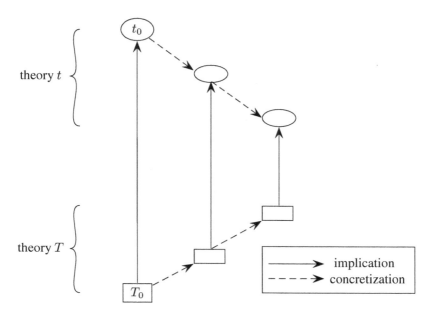

Figure 1. Essentially compatible idealizational theories

According to (T_0), there are some conditions where the following dependence holds:

$$D(x) = f(N(x))$$

But given coordinating definitions, this dependence in t amounts to:

$$C(x) = k_1(m_2(x)),$$

which is but a corrective function, showing how the secondary factor m_2 influences the investigated magnitude. This is no way near the dependence necessary for bringing about the reduction of t to T, i.e. $C(x) = g(M(x))$. The more so that we also need to take into account the conditions under which the dependences are to obtain. Given the above coordinating definitions, from T we will be able to derive:

$$(T_0')(x)[M(x) = 0 \wedge m_1(x) = 0 \rightarrow C(x) = h(m_2(x))]$$
$$(T_1')(x)[M(x) \neq 0 \wedge m_1(x) = 0 \rightarrow C(x) = h'(m_2(x), M(x))]$$
$$(T_2')(x)[M(x) \neq 0 \wedge m_1(x) \neq 0 \rightarrow C(x) = h''(m_2(x), M(x), m_1(x))]$$

While it is noteworthy that we will be able to derive the factual statement of theory t, the derivation condition is not satisfied for the idealizational statements of the

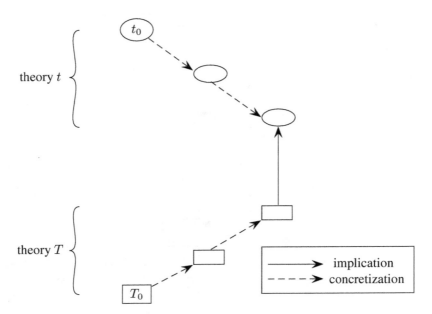

Figure 2. Essentially compatible idealizational theories

theory. The relationship between essentially incompatible idealizational theories is shown in Figure 2.

It is important to emphasize, on the one hand, that it is possible to derive the factual statements of essentially incompatible theories and, on the other, that it is impossible to derive their idealizational statements. The fact that it is impossible to derive the idealizational statements together with the idealizational law, which registers the regularity obtaining between the investigated and the principal factor, is intrinsically related to the fact that it is impossible to preserve the explanatory structure of the old theory in the new theory. According to Nowak [1977; 1980], explanation consists in showing first how the principal factor affects the investigated magnitude in the absence of secondary factors, and then slowly modifying that relationship to take into account the secondary factors so as to yield the empirical relationship exemplified in a given case. Given that the idealizational statements of the theories will not be preserved, the peak of the explanatory process will not be preserved either. This tallies nicely with the thought that the failure of reduction goes hand in hand with the failure of explanation.

However, it is also important to emphasize that it is possible to derive the factual statements of essentially incompatible theories. This means that the relation between the theories is non-accidental. They allow for there to be a convergence between the theories on detailed and relatively particularized generalizations (what Nancy Cartwright [1983] has called phenomenological laws). In other words, these theories explain the same phenomena but in different ways.

7   Objections Reconsidered

The Threat of Type-Epiphenomenalism. We will remember that the objection begins with the thesis that all causal relations among events obtain in virtue of their physical properties. The objection arises when—in order to rescue anti-reductionism—one rejects the type-type identity theory and is consequently unable to uphold the thesis that causal relations obtain in virtue of their mental properties. The proposal presented above shows how it is possible to accept non-reductionism without denying type-type identities. We can thus reject what was blocking the intuitive view that causal relations obtain not only in virtue of the physical but also in virtue of the mental properties. Essential incompatibility between idealizational theories is based on type-type identities between the factors of the theories in question without, however, preserving their essential orderings. This guarantees differences in the idealizational statements of both theories, preserving not only a token-token but also a type-type convergence on what but not how the theories explain.

The Threat of Dualism. We will remember that Kim's dilemma starts with a choice: either one chooses the view that mental causal powers are dependent on physical causal powers, which seems to commit one to reductive materialism, or one chooses the view that mental causal powers are independent of physical causal powers, which seems to commit one to dualism. There seems to be no room for a non-reductive version of materialism in this scenario.

Again, our discussion shows that there is a third way. We can accept for the sake of the argument that the causal powers of mental properties depend on the causal powers of physical properties. This does not yet prejudge the fact that psychological theories will be reducible to physical theories because the causal powers of properties do not determine the essential ordering of factors.

8   Conclusion

We have seen how the debate between reductionism and anti-reductionism can be enriched by taking into account developments in the philosophy of science. Using the framework of the Idealizational Conception of Science, I have shown that it is possible to uphold the anti-reductionist view while at the same time avoiding at least some of the objections that have been levelled against it. I have argued that taking idealization into account allows one to drive a wedge between type-type identity and reductionism. One way to claim that one theory does not reduce to another is to claim that although there is an identity among tokens, there is no identity among types, and so that there is no reduction of any generalizations. The other way to claim that one theory does not reduce to another is to claim that although there are type-identities among the factors of the theories, there are still no correspondences between the ways in which these factors figure into the hierarchy of laws of the theories in question. And while there may be correlations between some distant concretizations at the bottom of the hierarchies, this does not amount to the reduction of one theory to the other, since there is no correlation between the most general laws at the top of the hierarchies. In short, it is no longer imperative to postulate only token-token identities to save us from reductionism.

The conclusion of this paper is only apparently anti-reductionist. When two idealizational theories stand in the relation of essential incompatibility, it would be impossible to reduce either to the other. Whether such a relation takes place among any theories—and psychology and neurophysiology in particular—is and remains an open question. The task of philosophy is to understand the very nature of reduction and autonomy. Whether reductionism or anti-reductionism are the correct views to hold will, however, be decided by those who are best equipped to make this decision——by scientists in the near and, mostly likely, the distant future.

Acknowledgements

I have benefited greatly from comments and suggestions made by a number of people. I want to mention in particular: Jeremy Butterfield, Theo Kuipers, Krzysztof Łastowski, Leszek Nowak, Włodzimierz Rabinowicz and Marcel Weber.

BIBLIOGRAPHY

[Block, 1997] N. Block. Anti-Reductionism Slaps Back. *Philosophical Perspectives*, **11**, 107–132, 1997.

[Brzeziński *et al.*, 1990a] J. Brzeziński, F. Coniglione, T.A.F. Kuipers and L. Nowak, eds. *Idealization I: General Problems*. Amsterdam-Atlanta, GA: Rodopi, 1990.

[Brzeziński *et al.*, 1990b] J. Brzeziński, F. Coniglione, T.A.F. Kuipers and L. Nowak, eds. *Idealization II: Forms and Applications. Poznań Studies in the Philosophy of the Sciences and Humanities*, vol 17. Amsterdam-Atlanta, GA: Rodopi, 1990.

[Brzeziński and Nowak, 1992] J. Brzeziński and L. Nowak, eds. *Idealization III: Approximation and Truth. Poznań Studies in the Philosophy of the Sciences and Humanities*, vol. 25. Amsterdam-Atlanta, GA: Rodophi, 1992.

[Cartwright, 1983] N. Cartwright. *How the Laws of Physics Lie*. Oxford: Clarendon Press, 1983.

[Cartwright, 1989] N. Cartwright. *Capacities and their Measurement*. Oxford: Oxford University Press, 1989.

[Davidson, 1970] D. Davidson. Mental Events. Reprinted in *Essays on Actions and Events*, pp. 207–228. Oxford: Clarendon Press, 1970.

[Davidson, 1993] D. Davidson. Thinking Causes. In J. Heil and A. Mele, eds. *Mental Causation*, pp. 3–18. Oxford: Clarendon Press, 1993.

[Fodor, 1974] J. Fodor. Special Sciences, or the Disunity of Science as a Working Hypothesis. Reprinted in N. Block, *Readings in Philosophy of Psychology*, pp. 120–133. Cambridge, MA: Harvard University Press, 1980.

[Fodor, 1997] J. Fodor. Special Sciences: Still Autonomous after All these Years. *Philosophical Perspectives*, **11**, 149–163, 1997.

[Gaa, 1975] J. Gaa. The Replacement of Scientific Theories: Reduction and Explication. *Philosophy of Science*, **42**, 349–372, 1975.

[Honderich, 1982] T. Honderich. The Argument for Anomalous Monism. *Analysis*, **42**, 59–64, 1982.

[Kim, 1989] J. Kim. The Myth of Nonreductive Materialism. Reprinted in [Kim, 1993b, pp. 265–284].

[Kim, 1992] J. Kim. Multiple Realization and the Metaphysics of Reduction. *Philosophy and Phenomenological Research*, **52**, 1–26, 1992.

[Kim, 1993a] J. Kim. Can Supervenience and 'Non-Strict Laws' Save Anomalous Monism? In J. Heil, and A. Mele, eds. *Mental Causation*, pp. 19–26. Oxford: Clarendon Press, 1993.

[Kim, 1993b] J. Kim. *Supervenience and Mind. Selected Philosophical Essays*. Cambridge: Cambridge University Press, 1993.

[Kim, 1998] J. Kim. *Mind in a Physical World. An Essay on the Mind-Body Problem and Mental Causation*. Cambridge, MA: The MIT Press, 1998.

[Krajewski, 1977] W. Krajewski. *The Principle of Correspondence and the Growth of Knowledge*. Dordrecht-Boston: Reidel, 1977.

[Kuhn, 1962/1970] T. Kuhn. *The Structure of Scientific Revolutions*. 2nd edition. Chicago: The University of Chicago Press, 1970.

[Kuipers, 1982] T. A. F. Kuipers. The Reduction of the Phenomenological to Kinetic Thermostatics. *Philosophy of Science*, **49**, 107–119, 1982.

[Kuipers, 1985] T. A. F. Kuipers. The Paradigm of Concretization: The Law of van der Waals. In J. Brzeziński, ed., *Consciousness: Methodological and Psychological Approaches*, pp. 185–199. *Poznań Studies in the Philosophy of the Sciences and the Humanities*, vol. 8, pp. 185–199. Amsterdam: Rodopi, 1985.

[Kuipers, 1990] T. A. F. Kuipers. Reduction of Laws and Concepts. In [Brzeziński *et al.*, 1990a, pp. 241–276].

[LePore and McLaughlin, 1985] E. LePore and B. P. McLaughlin, eds. *Actions and Events. Perspectives on the Philosophy of Donald Davidson*. Oxford: Basil Blackwell, 1985.

[Lewis, 1966] D. Lewis. An Argument for the Identity Theory. *Journal of Philosophy*, **63**, 17–25, 1966.

[Lewis, 1972] D. Lewis. Psychophysical and Theoretical Identifications. *Australasian Journal of Philosophy*, **50**, 249–258, 1972.

[Machowski, 1990] A. Machowski. Significance: An Attempt at Variational Interpretation. In [Brzeziński *et al.*, 1990a, pp. 23–31].

[McDowell, 1985] J. McDowell. Functionalism and Anomalous Monism. In [LePore and McLaughlin, 1985, pp. 387–398].

[McLaughlin, 1989] B. P. McLaughlin. Type Epiphenomenalism, Type Dualism, and the Causal Priority of the Physical. *Philosophical Perspectives*, **3**, 109–136, 1989.

[Nagel, 1961] E. Nagel. *The Structure of Science*. New York: Harcourt, Brace and World, 1961.

[Nowak, 1971] L. Nowak. *U podstaw marksowskiej metodologii nauk*, Warszawa: PWN, 1971.

[Nowak, 1977] L. Nowak. *Wstęp do idealizacyjnej teorii nauki*. Warszawa: PWN, 1977.

[Nowak, 1980] L. Nowak. *The Structure of Idealization*. Dordrecht/Boston: Reidel, 1980.

[Nowak, 1989] L. Nowak. On the (Idealizational) Structure of Economic Theories. *Erkenntnis*, **30**, 225–246, 1989.

[Nowak, 1992] L. Nowak. The idealizational approach to science: A survey. In [Brzeziński and Nowak, 1992, pp. 9–63].

[Nowak, 2000] L. Nowak. The Idealizational Approach to Science: A New Survey. In [Nowak and Nowakowa, 2000, pp. 109–184].

[Nowak and Nowakowa, 2000] L. Nowak and I. Nowakowa. *The Richness of Idealizations. Poznań Studies in the Philosophy of the Sciences and the Humanities*, vol. 69. Amsterdam-Atlanta: Rodopi, 2000.

[Nowakowa, 1975] I. Nowakowa. *Dialektyczna korespondencja a rozwój nauki*. Warszawa-Poznań: PWN, 1975.

[Nowakowa, 1994] I. Nowakowa. *The Dynamics of Idealizations*. Amsterdam-Atlanta, GA: Rodopi, 1994.

[Paprzycka, 1990] K. Paprzycka. Reduction and Correspondence in the Idealizational Approach to Science. In [Brzeziński *et al.*, 1990a, pp. 277–286].

[Paprzycka, 1999] K. Paprzycka. Socrates Meets Carnap: Explication in the Theaetetus. *Logical Analysis and History of Philosophy*, **2**, 87–108, 1999.

[Paprzycka, forthcoming] K. Paprzycka. *O możliwości anty-redukcjonizmu (On the Possibility of Anti-Reductionism), forthcoming.*

[Paprzycki and Paprzycka, 1992] M. Paprzycki and K. Paprzycka. Accuracy, Essentiality, Idealization. In [Brzeziński and Nowak, 1992, pp. 255–265].

[Putnam, 1967] H. Putnam. The Nature of Mental States. Reprinted in *Mind, Language and Reality: Philosophical Papers*, vol. II, pp. 429–440. Cambridge: Cambridge University Press, 1975.

[Salmon, 1984] W. C. Salmon. *Scientific Explanation and the Causal Structure of the World,* Princeton: Princeton University Press, 1984.

[Salmon, 1989] W. C. Salmon. Four Decades of Scientific Explanation. In P. Kitcher and W.C. Salmon, eds, *Scientific Explanation. Minnesota Studies in the Philosophy of Science*, Vol. XIII, pp. 3–219. Minneapolis: University of Minnesota Press, 1989.

[Schaffner, 1967] K. F. Schaffner. Approaches to Reduction. *Philosophy of Science*, **34**, 137–147, 1967.

[Soutland, 1976] F. Stoutland. The Causation of Behavior. In J. Hintikka, ed., *Essays on Wittgenstein in Honor of G.H. von Wright. Acta Philosophica Fennica*, **28**, 286–325. Amsterdam: North-Holland, 1976.

[Stoutland, 1985] F. Stoutland. Davidson on Intentional Behaviour. In [LePore and McLaughlin, 1985, pp. 44–59].

Katarzyna Paprzycka

Instytut Filozofii

Szkoła Wyższa Psychologii Społecznej (SWPS)

ul. Chodakowska 19/31

03-815 Warszawa, Poland

Email: Katarzyna.Paprzycka@swps.edu.pl

On the Epistemic Significance of Type-Type Identities
A Comment on Katarzyna Papryzcka

MARCEL WEBER

In the philosophy of mind, as well as in the general reductionism debate in the philosophy of science, it was generally assumed that a type-type identity theory relating two domains such as the mental and physical implies that theories from one domain would be fully reducible to theories from the other domain. In other words, type-type identities between two domains are universally thought to be *sufficient* for reduction.[1] To my knowledge, Katarzyna Paprzycka is the first to challenge this assumption. Her paper raises the intriguing possibility that even though there is a one-one mapping from types of mental states into types of physical states, there exist true or approximately true theories about the mental domain that are not reducible to true or approximately theories about the physical domain. The same could be true for other scientific theories, perhaps in the special sciences. According to Paprzycka, this possibility is generated by the idealized nature of some scientific theories. New insights from the philosophy of science concerning idealization have, in her view, hitherto unappreciated consequences for the philosophy of mind, and probably also for the philosophy of special sciences other than psychology.

As it is customary for commentators, I am going to challenge Paprzycka's conclusions. In order to be brief, I would like to point out just two problems in her line of argument, even though I think that there are more. For instance, her arguments as presented apply only to theories that are mathematically formulated, which is typically not the case in psychology and in neuroscience. But this problem is philosophically uninteresting. I shall instead focus on two problems that bring out some philosophical puzzles concerning the nature of identity and the deeper reasons why so many people might have thought that identities imply reduction.

The first problem concerns the argument from idealization and type-type identity to irreducibility. The second problem concerns the relevance of the argument for the philosophy of mind as well as the philosophy of special sciences. I will argue that even if the argument as presented were watertight—which it is not—it

---

[1] By contrast, there was some debate as to whether such type-type identities are also *necessary* for reduction, as Causey [1977] argued (cf. [Weber, 1996]).

Laws and Models in Science, 131–134.

would not matter for metaphysics, nor would it matter for the philosophy of special sciences. However, the reasons for this are far from being trivial, and we would have missed them if it weren't for Katarzyna Paprzycka's stimulating paper.

The first difficulty I will discuss concerns the concept of type-type identity as it is used in Paprzycka's paper. The standard sense of identity that underlies the debates in the philosophy of mind is that of a *synthetic* but possibly *necessary* (according to [Kripke, 1971]) relation between a set of things or states as they appear under one description, and the very same set of things or states as they appear under another description. For example, the state of containing heat is really *the same state* as that of being composed of atoms and molecules that move about in a random fashion. Heat is type-identical with molecular motion. It is the very same state, but viewed under two different descriptions, namely, a macroscopic and a microscopic description. Similarly, the class of light beams is type-identical with the class of electromagnetic waves of a certain wavelength, just viewed under two different descriptions (one phenomenological, the other electrodynamic). Thus, the nature of these identity relations is such that they relate two sets of things or states that are *not*, in fact, two sets of things or states, but just one an the same instead. Their distinctness is an illusion.

It is this kind of an identity relation that underlies the mind-body identity theory in its most sophisticated versions. Now, let us consider what kind of identity relation Paprzycka talks about in her paper.

Paprzycka presents us with abstract forms of scientific theories that state a law-like functional dependence of some magnitude $C(x)$ or $D(x)$ from a set of independent variables $M(x)$, $m_i(x)$, $N(x)$, $n_i(x)$. Each theory consists of a *series* of laws where the first member contains only one term expressing what is called the 'principal factor', while the subsequent members contain an increasing number of additional terms expressing factors that are less essential. The more additional terms a law contains, the less idealized it is.

Now, Paprzycka assumes that it be 'discovered' that the dependent magnitudes of two theories are equal, i.e., $C(x) = D(x)$, and that the independent variables are connected by 'coordinating definitions'. She distinguishes two cases: In the compatible case, the coordinating definitions respect the ordering of the factors with respect to essentiality, in the incompatible case, it does not. In the incompatible case, it will thus not be possible to derive all the different concretisations of the theory; at best, some specific values ('factual statements') are derivable. So far, I can find no fault in Paprzycka's reasoning.

What I fail to see, however, is how this mathematical scenario instantiates a type-type identity theory of the kind that has been envisioned by philosophers of mind (and some philosophers of science). True, in both the compatible and incompatible cases, the magnitudes are equal, and so are some of the other terms. But equality does not an *identity* make. The Jones family's annual expenditure on fine art may equal the Smith family's annual income, without this being the same money. In order for a type-type identity to hold, the terms in question must refer to exactly the same set of things or states. But if we have two theories that refer to exactly the same set of things or states, how can the incompatible case arise in the first place? The basic intuition behind the argument seems to be that two idealized theories identify the same factors as relevant for some theoretical magnitude, but rank them differently with respect to essentiality. But if the coordinated terms are not merely *equal* in magnitude, but express the *very same factors*, why should their essential ordering differ in the first place?

A possible defence is this: Both theories are not strictly true; they are *idealized* and thus at best approximately true. Thus, both theories are off the mark to some extent. They might both err with respect to the true essentiality of the different factors, but they could err in different ways. This is possible even if the factors that are represented in the two theories *are*, in fact, one and the same (i.e., type-identical).

But this defence pushes the source of the irreducibility into the *falsity* or, in other words, the *defective* parts of both theories involved. It therefore describes a situation in which scientists are in no position to assert any kind of type-type identity relation because they are faced with two incompatible theories that are both known to be defective.

These considerations suggest that Paprzycka's argument succeeds only by violating the spirit of a type-type identity theory and replacing it by a weaker relation that merely states the equality of some theoretical magnitudes. If the identity theory is taken seriously, the kind of incompatibility envisioned by Paprzycka is unlikely to arise. If it does arise, this indicates that both theories involved are off the mark, which undermines the identity theory itself.

Now I come to my second point. Paprzycka's goal is to use some recent advances in the philosophy of science, namely a theory of scientific idealization, in order make a substantial point about mind-brain reductionism. While there is nothing in principle wrong with this approach, it must take into consideration some *other* developments in the philosophy of science as well. This immediately leads to the recognition that Paprzycka's argument is based on a particular account of reduction, namely Ernest Nagel's derivational account [Nagel, 1961]. According to this account, a reduction is always a derivation of the laws of some higher-level theory from the laws of a more fundamental theory with the help of bridge principles. But the debate on reduction has also moved on since Nagel's famous and important 1961 book. First, it was suggested that Nagel's bridge principles must be replaced by a much stronger relation, namely an *a posteriori* and possibly necessary type-type identity relation [Causey, 1977]. This is a far cry from Nagel's bridge principles. But Paprzycka's "coordinating definitions" are basically Nagel-bridge principles, and I have shown that her argument fails to sting if these bridge principles are replaced by the genuine identity relations envisioned by philosophers of mind.

Second, the idea that reduction is always a *derivation* of some laws has long been abandoned. In the philosophy of biology, for example, it was shown that the laws of classical genetics are not derivable from the principles of molecular biology [Hull, 1974]. However, non-derivational accounts of reduction can accommodate the case [Waters, 1990]. If we had molecular explanations of mental phenomena that could match the kind of understanding of biological phenomena provided by molecular genetics, I doubt that metaphysicists would get excited and exclaim: Look! But psychological theories are not strictly *derivable* from neurophysiology.[2] Metaphysicists (and I do not mean this term in a derogatory sense) are more likely to object that even if we knew the neural correlates of all possible types of mental states, this would not furnish an *explanation* for some of their qualitative aspects. In other words, if a type-type identity theory about the mind/brain could be *shown* to be true, who would care what is and what is not derivable from some idealized theory? The salient question would rather be what exactly the identity theory *explains*. But mathematical derivability is neither necessary nor sufficient for explanation.

Thus, Paprzycka argument can only work against an unreasonably strong account of reduction, namely one that requires derivability.

BIBLIOGRAPHY

[Causey, 1977] R. Causey. *Unity of Science*. Dordrecht: Reidel, 1977.
[Hull, 1974] D. Hull. *Philosophy of Biological Science*. Englewood Cliffs: Prentice Hall, 1974.
[Kim, 1998] J. Kim. *Mind in a Physical World: An Essay on the Mind-Body Problem and Mental Causation*. Cambridge, MA: MIT Press, 1998.
[Kripke, 1971] S. A. Kripke. Identity and Necessity. In *Identity and Individuation*, edited by M. K. Munitz, pp. 135–164. New York: New York University Press, 1971.
[Nagel, 1961] E. Nagel. *The Structure of Science. Problems in the Logic of Scientific Explanation*. London: Routledge and Kegan Paul, 1961.
[Waters, 1990] C. K. Waters. Why the Anti-Reductionist Consensus Won't Survive the Case of Classical Mendelian Genetics. In *PSA 1990*, edited by East Lansing: Philosophy of Science Association, pp. 125–139, 1990.
[Weber, 1996] M. Weber. Fitness Made Physical: The Supervenience of Biological Concepts Revisited. *Philosophy of Science*, **63**, 411–431, 1996.

Marcel Weber

Centre for Philosophy and Ethics of Science

University of Hannover

Im Moore 21

30167 Hannover, Germany

Email: weber@ww.uni-hannover.de

---

[2]Kim [1998] presents a model for mind-brain reduction that does not involve the derivation of laws as was suggested by Nagel.

Response to Marcel Weber's Comment

KATARZYNA PAPRZYCKA

I want to thank Marcel Weber for his incisive and constructive criticism of my paper and I would like to use the space below to address the major worry he raises about the very intelligibility of the proposal defended in the paper. Before I turn to it, I will make a brief comment about his second major criticism, viz. that I use a derivational conception of reduction. I simply want to acknowledge that I do so in this paper – to a large extent for the purposes of an argument. I quite agree that we may want to talk about reductive relations between theories that are not formulated in ways that would allow for a derivation and I by no means think that derivability is essential to reduction. The larger project [Paprzycka, forthcoming], of which this paper is a part, attempts to cast bridges between Salmon's [1984; 1989] and Nowak's [1980] theories of science. As is well known, Salmon has been hostile to requiring any derivability conditions in the account of explanation and probably would be in an account of reduction. Still, I believe that it is not necessary to take such a radical stance, and while we should learn the lesson Salmon teaches us—that relevance matters—this is no reason to think that 'controlled' derivability relations cannot be understood as showing us something important. In the Idealizational Conception of Science, derivability occurs only against the background of essentiality relations. Without going into too much detail, let us only remind ourselves of the footnote 33 problem to which pure derivability conceptions of reduction are subject, i.e. the problem of generating spurious reductions (derivations) of a theory t from a conjunction of t and an arbitrarily chosen other theory T_X. What blocks spurious derivations in the Idealizational Conception of Science is the fact that the derivations are to manifest deeper essentiality relations—as long as theory T_X does not investigate the same factor as t, and as long as its space of essential factors does not at least partially overlap with the space of essential factors of t, we cannot even begin to think that a reduction is involved.

Let us then turn to the problem of having the identity-theory cake and eating it with an anti-reductionist pudding. Since I *do* mean factor identities to be factor identities and not just equalities among the values that different factors take, Weber points out that the following objection arises: If it turns out that the factors of two true theories can be identified with each other in such a way that the essential structure of the factors will not be preserved under the identification, why should

Laws and Models in Science, 135–138.

we not conclude that one of the theories is simply wrong about the essential ordering of factors rather than thinking that they are equally good but irreducible to one
another?

Two points must be appreciated before we can see how one could defend the
intelligibility of suggestion that irreducibility can be based on essential incompatibility. (1) We have been assuming thus far that the factors of the two theories can be identified one-by-one with one another. It is highly unlikely, however, that the factors of mature psychology could be identified on a one-by-one
basis with the factors of physics or neurophysiology. We will thus waive this
assumption allowing for the possibility that higher-level factors be more complex constructs composed of lower-level factors. (2) The second point concerns
the question how to understand the notion of essentiality. On one (the accuracy) interpretation of the degrees of essentiality [Paprzycki and Paprzycka, 1992;
Paprzycka, forthcoming], scientists order factors depending on the degree of accuracy that the inclusion of a factor affords.

With these thoughts in mind, let us consider the following schematic example.
Let us assume that we are dealing with two true independently proposed theories
that explain apparently different phenomena. Let us assume that theory t proposes
the following essential structure of C:

$$S_C : \quad M$$
$$M, m_1$$
$$M, m_1, m_2$$

The essential structure of factor D, which is the investigated factor of theory T
looks thus:

$$S_D : \quad N$$
$$N, n_1$$
$$N, n_1, n_2$$
$$N, n_1, n_2, n_3$$
$$N, n_1, n_2, n_3, n_4$$
$$N, n_1, n_2, n_3, n_4, n_5$$
$$N, n_1, n_2, n_3, n_4, n_5, n_6$$

As it turns out, the two theories have been in fact investigating the same factors:

$$C = D$$

and that the remaining factors of theory t can be identified as constructs of the
factors of theory T, so that the following identities hold:

$$M = n_1 \wedge n_2 \wedge n_4$$
$$m_1 = N \wedge n_5$$
$$m_2 = n_3 \wedge n_6$$

The relation between the factors is demonstrated by Fig. 1, where in particular the degree of essentiality of factors is shown.

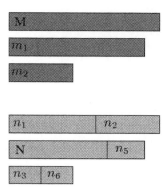

Figure 1. The relation between the factors of two essentially incompatible theories t and T. The areas represent the accuracy afforded by the inclusion of a given factor. Further explanation in text

The diagram makes clear why a theory that uses factors M, m_1, m_2 would consider them to be essential to the investigated magnitude in that order. The fact that it is possible to identify those factors with $n_1 \wedge n_2 \wedge n_4$, and $N \wedge n_5$, and $n_3 \wedge n_6$, respectively, does not provide a reason to undermine their ordering—quite to the contrary it actually allows us to vindicate it. A theory that uses factors M, m_1, m_2 must use them in this—and no other—order. In this way, the assumption that factors of theory t are complex constructions of factors of theory T together with the accuracy interpretation of essentiality provide a justification why the essential structure of theory T need not lead to a change in the essential structure of theory t. This suffices to establish the intelligibility of the suggestion that essentially incompatible theories will be irreducible to each other.

BIBLIOGRAPHY

[Nowak, 1980] L. Nowak. *The Structure of Idealization*. Dordrecht/Boston: Reidel, 1980.
[Paprzycka, forthcoming] K. Paprzycka. *O możliwości anty-redukcjonizmu (On the Possibility of Anti-Reductionism)*, forthcoming.
[Paprzycki and Paprzycka, 1992] M. Paprzycki and K. Paprzycka. Accuracy, Essentiality, Idealization. In J. Brzeziński and L. Nowak, *Idealization III: Approximation and Truth. Poznań Studies in the Philosophy of the Sciences and the Humanities*, vol. 25. Amsterdam/Atlanta, GA: Rodopi, 1992.
[Salmon, 1984] W. C. Salmon. *Scientific Explanation and the Causal Structure of the World*. Princeton: Princeton University Press, 1984.

[Salmon, 1989] W. C. Salmon. Four Decades of Scientific Explanation. In P. Kitcher and W.C. Salmon, eds, *Scientific Explanation. Minnesota Studies in the Philosophy of Science*, vol. XIII, pp. 3–219. Minneapolis: University of Minnesota Press, 1989.

Katarzyna Paprzycka

Instytut Filozofii

Szkoła Wyższa Psychologii Społecznej (SWPS)

ul. Chodakowska 19/31

03-815 Warszawa, Poland

Email: Katarzyna.Paprzycka@swps.edu.pl

Idealisation and Mathematisation in Cassirer's Critical Idealism

Thomas Mormann

1 Idealist Philosophy of Science?

In Anglo-American philosophy there is a strong conviction that idealism on the one hand, and science and serious philosophy of science on the other hand, do not go well together. Often, idealism plays the role of a strawman to whom all the vices are attributed that one wants to criticise. As a particular savage example of this kind of anti-idealism let me mention Israel Scheffler's characterisation of Thomas Kuhn as an irresponsible and even immoral idealist:

> The current attacks (of Kuhn, T. M.) challenge ...the very opposition between science and speculative idealism, from which scientifically minded philosophies have sprung. The attacks threaten further the underlying moral motivation of these philosophies, their upholding of the ideal of responsibility in the sphere of belief as against willfullness, authoritarianism, and inertia. The issues are fundamental, indeed more fundamental than is generally realized, precisely because a powerful moral vision has implicitly been called into question. [Scheffler, 1967, pp. 7–8]

Perhaps Scheffler's attack can be seen as a remote rehearsal of Moore's and Russell's anti-idealist revolt against British idealism around the turn of the last century. Be that as it may, in the Anglo-American philosophical scene the opinion is wide-spread that all variants of idealism subscribe to the doctrine that 'reality is fundamentally mental'. As typical one may take Haack's characterisation according to which '[A]n idealist holds that everything there is, is mental: that the world is a construction out of our ideas', [Haack, 2002, p. 70]. A bit more specifically Russell maintained that idealists believe that all propositions are of the subject-predicate form, and therefore idealists do not appreciate the merits of modern relational logic. As will be shown in the following, Cassirer may serve as a brilliant counter-example to the claims of Russell and Haack: it is difficult to find a philosopher who praised the achievements of relational logic more ardently than Cassirer.

Laws and Models in Science, 139–159.

Despite, or perhaps just because of its oversimplified character, conceptions of idealism like Haack's are rather common ones. Consequently, in contemporary analytic philosophy of science, idealist positions are rarely mentioned. For instance, in most books that intend to give a history of philosophy of science in the twentieth century neither Cassirer nor any other idealist is mentioned at all.

If idealism were what its caricature maintains it to be this would be alright, but actually things are more complicated. To show why, one may start with a bit of history of philosophy of science: nobody will deny the importance of Logical Empiricism for philosophy of science of the 20th century. Some authors even contend that contemporary philosophy of science is to be conceived of as a successor discipline of Logical Empiricism. As has been shown by detailed studies by Coffa, Friedman, Richardson and others, most logical empiricists started their philosophical careers as neo-Kantian idealists. A case in question is Carnap: not only that Carnap began as a neo-Kantian idealist, even worse, in *The Logical Structure of the World* [Carnap, 1928] he openly confessed to have sympathies for idealist doctrines. For instance, he characterized the constitutional theory which may safely be called the core of the *Aufbau* program, as deeply influenced by transcendental idealism:

> The merits of having discovered the necessary basis of the constitutional system thereby belongs to two entirely different . . . philosophical tendencies. *Positivism* has stressed that the sole material for cognition lies in the undigested experimental given . . . Transcendental idealism, however, especially of neo-Kantian tendency (Rickert, Cassirer, Bauch), has rightly emphasised that these elements do not suffice; order-posits must be added, our *basic relations*. [Carnap, 1928, Section 75]

Thus, in one of the key works of logical empiricist philosophy of science we find a large dose of idealism. This and other evidences show that there is something in idealism that does not go away as easily as many may wish. Summarising, then, I'd like to assent to Crispin Wright who put the problem with idealism in the following way:

> For all the vilification and caricature which its critics have meted out over the years, the idealist tradition in philosophy has proved sufficiently durable to encourage the belief that, at least locally, there are insights for which it is striving, but for which—its persistently controversial character suggests—we have yet to find definitive means of expressions. [Wright, 1992, p. 3]

After these introductory remarks let me come to my point, the role of idealisation and mathematisation in Cassirer's 'critical idealism'. More precisely I propose to reconsider a central thesis of Cassirer's that deals with the role of idealisation and mathematisation in the sciences, in particular, in physics. This is done not only for an interest in history of philosophy of science. Rather, I contend that Cassirer's thesis might be of some interest for the contemporary agenda of philosophy of science dealing with the roles of idealisation, mathematisation, laws, and models. Cassirer put forward his thesis in the early paper *Kant und die Moderne Mathematik* ([Cassirer, 1907], KUMM in the following). The title of this bulky article is a bit misleading, the issue is not so much Kant but the problem of how neo-Kantian philosophy of science should assess the recent developments of logic and mathematics, in particular the growing importance of the theory of relations for logic, mathematics and philosophy in general. KUMM may be considered as the programmatic precursor of Cassirer's first opus magnum *Substance and Function*, [?].

It goes without saying that the thesis of Cassirer's I want to discuss in the following is not an isolated assertion but embedded in a rather complex theoretical context, to wit, his theory of the formation of scientific concepts. Indeed, according to him, philosophy of science is to be conceived as the theory of the formation of scientific concepts. Since this paper is not the appropriate place to develop the essentials of Cassirer's account in an orderly manner, the following six theses may be distilled from *Substance and Function* and later works (cf. Cassirer [1929; 1937]).

1. Scientific knowledge does not cognize objects as ready-made entities. Rather, knowledge is organized objectually in the sense that in the continuous stream of experience invariant relations are fixated.

2. The unity of a concept is not to be found in a fixed group of properties, but in the rule, which lawfully represents the mere diversity as a sequence of elements. The meaning of a concept depends on the system of concepts in which it occurs. It is not completely determined by one single system, but rather by the continuous series of systems unfolding in the course of history. Scientific knowledge is a 'fact in becoming' ('Werdefaktum').

3. Scientific concepts and conceptual systems do not yield pictures of reality, rather, concepts provide guide lines for the conceptualisation of the world. The fundamental concepts of theoretical physics are blueprints for possible experiences.

4. Factual and theoretical components of scientific knowledge cannot be neatly separated. In a scientific theory 'real' and 'non-real' components are inextricably interwoven. Not a single concept is confronted with reality but a whole system of concepts.

5. Our experience is always conceptually structured. There is no non-conceptually structured 'given'. Rather, the 'given' is an artifact of a bad metaphysics.

6. The concepts of mathematics and the concepts of the empirical sciences are of the same kind.

In the following, I'd like to concentrate on (6). Implicitly, however, we have to deal with the other theses of Cassirer's as well. As a start, it may be expedient to quote (6) more fully as follows:

> What 'critical idealism' seeks and what it must demand is a *logic of objective knowledge* (gegenständliche Erkenntnis). Only when we have understood that the *same foundational syntheses* (Grundsynthesen) on which logic and mathematics rest also govern the scientific construction of experiential knowledge, that they first make it possible for us to speak of a strict, lawful ordering among appearances and therewith of their objective meaning: only then the true justification of the principles is attained. [Cassirer, 1907, p. 44]

This thesis will be referred to as the 'sameness thesis' (henceforth ST). I'd like to contend that ST lies at the heart of the 'critical idealist' philosophy of science Cassirer first presented in *Substance and Function* and later elaborated throughout his entire philosophical career (cf. Cassirer [1929; 1937]).

The outline of this paper is as follows: in Section 2 some preliminary comments on ST are put forward in order to forestall some unnecessary misunderstandings. In order to set the stage for a proper assessment of ST in Section 3 we reconsider some paradigmatic examples of the introduction of ideal elements that may be considered as point of departure for Cassirer's account. In Section 4 we deal with idealisation in mathematical physics in order to render plausible ST for the realms of mathematics and physics. In Section 5 we conclude with some general remarks on the place of Cassirer's 'critical idealism' in the landscape of 20th century's philosophy of science.

2 Some Preliminary Comments on ST

At first glance, one may be tempted to read the 'sameness thesis' ST as a sort of vulgar idealism which *identifies* mathematics and physics. This would be a misunderstanding. According to Cassirer, philosophy as philosophy of science has to concentrate *neither on mathematics*, as an ideal science, *nor on physics* as an empirical science, but rather:

> If one is allowed to express the relation between philosophy and science in a blunt and paradoxical way, one may say: The eye of philosophy must be directed neither on mathematics nor on physics; it is

to be directed solely on the connection of the two realms. [Cassirer, 1907, p. 48]

For Cassirer's philosophy of science the central point of reference is neither mathematics—as a science of ideal objects, nor physics—as a purely empirical science. Cassirer did *not* aim at the futile reduction of physics to mathematics or an identification of both. Rather, he was looking for a common root from which both physics *and* mathematics spring. This common root is identified as the idealising method of the introduction of ideal elements.

Today, when dealing with idealisation in science it is usually taken for granted that there is a strict separation between the mathematical and the physical realm. Implicitly it is assumed that within mathematics there is no place for idealisation. Mathematics already is on the ideal side, so to speak. Under this assumption, the problem of the idealisational character of scientific knowledge is said to be solely concerned with the problem of the role of idealisation in the empirical realm. For instance, Leszek Nowak and his school have set up a detailed classification of the various methods of idealisation, but they are concerned only with the various forms of idealisations in the empirical sciences. They never consider mathematics as a domain for which idealisation could be relevant (cf. [Nowakowa and Nowak, 2000]).

According to Cassirer such a theory of idealisation starts too late: for him, idealisation has a role in mathematics *and* in the empirical sciences. Hence, a theory of idealisation in science has to take into account both mathematics *and* the empirical sciences. If one wants to understand the role of idealisation in empirical science one should study how it works in mathematics and in empirical science. Moreover, one should not tackle this problem armed with 'philosophical' presuppositions of what are the philosophically correct methods of idealisation. The methods of idealisation should be studied empirically, so to speak, no philosophical intuition will give us the answer what the common foundational syntheses are on which logic, mathematics and empirical science are based. Rather, this has to be revealed by studying the history of science. For Cassirer this meant to study the history of the formation of scientific concepts. Hence, philosophy of science has to pay attention to the ongoing evolution of science, it has to investigate and explicate the formation of scientific concepts in the real history of science. [1] In a nutshell, then, *ST* contends:

---

[1] This entails a specific difficulty for contemporary philosophy of science, in particular philosophy of mathematics. Today, the latter is not too much interested in 'real mathematics' and its historical development, but rather in its so-called logical foundations. The more advanced topics of mathematics are assumed to contribute nothing to the philosophical understanding. In contrast, Cassirer was interested in contentful mathematics, since a basic assumption of his philosophy of science was that science and mathematics can only be understood by studying its real history. Hence, his answer of what are the common foundational syntheses requires something more than elementary mathematics.

> *The common foundational syntheses on which mathematical and phys-*
> *ical knowledge are based, are provided by the method of 'ideal ele-*
> *ments'.*

The point Cassirer wants to make is that mathematicians and physicists are *both* using the methods of ideal elements. This claim is in stark contrast with modern views according to which the mathematicians are, so to speak, already *inside* the sphere of ideal objects, while the physicists, when they are offering their idealized laws and models that do not hold good in the real world, somehow attempt to enter the ideal domain but, strictly speaking, never succeed in getting in.

ST claims that this dichotomy is misleading. Mathematics should not be characterized as a realm of ideal objects while physics is said to be confined to the crude and non-ideal empirical sphere. Principally, the domain of mathematics is not too different from that of physics. Also within mathematics a lot of idealising is necessary to formulate and prove interesting theorems in a sea of ephemeral phenomena. In other words, one has to provide appropriate settings [2] in order to be able to do some interesting work in mathematics. This involves some sort of idealising. In order to render plausible this thesis one has to explain what is the role of ideal elements in the realm of geometry, or where idealisation takes place in the realm of numbers. Cassirer gave detailed answers to these questions, but today most philosophers seem to ignore them. For instance, in a recent discussion on the influence of neo-Kantian idealism on Carnap's *Aufbau* Alan Richardson comes to the conclusion that Cassirer failed to say clearly what is to be understood by these foundational syntheses, and the allegedly 'common foundational syntheses' remained mere verbiage [Richardson, 1997]. I think that Richardson's verdict is too hasty. Although a comprehensive answer to what are the common foundational syntheses may be said to be missing in KUMM, a full-fledged answer can be found in *Substance and Function*.

This does not mean that this answer is still fully satisfying today. The reason is not that Cassirer hadn't offered a good account of the methodology of ideal elements. Quite the contrary. Since his days the methodology of ideal elements has made great progress and therefore his account needs an update. In order to show that Cassirer's account of the method of ideal elements is not obsolete for understanding the role of idealisation one has to go beyond elementary mathematics. Standard elementary mathematics is sanitised in such a way that the role of idealising in it is hidden from the eye. This should not be too surprising: for quite a long time the idealisational character of physical knowledge has been ignored by philosophy as well. Concentrating on some toy theories encapsulated in sentences like 'Copper expands when heated' or 'All swans are white' the indispensable role of idealisation hardly springs to the eye. Hence, with respect to matters of ideali-

---

[2]This expression has been coined by Wilson [1992].

sation, a parallelity between 'real' mathematics and 'real' physics arises: in both fields its role has been neglected by philosophy of science. Whilst philosophers of science have made some progress to get into contact with 'real physics', the analogous re-orientation towards 'real mathematics' in philosophy of mathematics is still in its beginnings.[3] In this respect, a reconsideration of Cassirer's philosophy of science that deliberately dealt with mathematics and physics might be helpful.[4]

3   Ideal Elements in Mathematics

Till the beginnings of the 19th century, a non-expert, for instance a philosopher, might have been justified to conceptualise the domain of geometry as an unalterable sphere of ideal objects such as ideal points, ideal lines etc. From that time onwards, however, it became more and more evident that Euclidean geometry was less than perfect and ideal. Seen from a mathematical perspective, it could be said to have certain conceptual defects which called for fixing. To formulate it in a somewhat paradoxical way: too many theorems one wanted to be true, turned out not to be true. Perhaps the simplest example was provided by projective geometry of the plane. From a mathematical point of view it had long been known that between points and lines there existed a certain useful *duality*: for a given theorem it was sometimes possible to obtain a new theorem by switching the terms 'point' and 'line': for instance, given the proposition that every two points determine a single line, the dual proposition was that every two lines determine a point by their intersection. Or, a triangle could be defined by its three vertices as well as by its three intersecting sides.[5]

Unfortunately, in Euclidean geometry a dual of a theorem was not always a theorem. For instance, although two points always determine a unique line, two lines not always determine a point since two parallels do not intersect. The method of ideal elements was to fix deficiencies of this kind. It introduced new 'ideal points' located on a new 'ideal line' that rendered the originally incomplete duality perfect.[6] This could be done in several ways. One method was to conceive an ideal

---

[3]To be sure, there are promising exceptions, for instance, Corfield's recent book 'Towards a Philosophy of Real Mathematics' [Corfield, 2003].

[4]Before we go on it should be noted that Cassirer's emphasis of the important role of ideal elements in mathematics is not a special feature of his 'critical idealism' or the neo-Kantianism of the Marburg school. In the 19th century, the issue of ideal elements was a common topic discussed by philosophers of mathematics and mathematicians alike (cf. [Wilson, 1992]). This continued till the first decades of the 20th century. After the rise of logicism and formalism the talk about ideal elements was no longer to be considered a serious theme in philosophy of mathematics. In mathematics itself, mathematicians continued to talk about them, but, as it seems, philosophers no longer were interested to listen.

[5]As a less trivial example one may mention the dual theorems of Pascal and Brianchon, see[Smith, 1959, pp. 331–336].

[6]For an insightful discussion of 'ideal elements' in the case of complex projective geometry, see [Wilson, 1992]. Here, I restrict myself to some brief remarks on the more elementary case of real projective geometry. An easily accessible, more detailed discussion can be found in[Torretti, 1978].

point as an equivalence class of parallel lines in such a way that any two parallel lines intersect at this ideal point. This sounded paradoxical, since the new 'point', being an equivalence class of lines, seemed to be larger than the 'line' it was to be a part of. Nevertheless, this method worked, and the new points and lines could be shown to fulfil the tasks they were designed for.[7]

When they were introduced, these ideal elements aroused some suspicion among the more conservative minded mathematicians. Ideal points and ordinary points had not the same ontological dignity, so to speak. Only gradually ideal points and ideal lines became citizens on a par with 'ordinary' points and lines. An important step on the road to full recognition was the construction of models of projective spaces. For instance, the points of the real projective plane could be conceptualised as the set of lines through the origin of the real 3-dimensional vector space, and correspondingly the lines of the projective plane could be conceived of as planes through the origin.[8] As a result of this and other developments, in the second half of the 19th century geometry was no longer considered as the investigation of an immutable domain of ideal objects but rather as an unfolding theory of generalised spatial structures defined by appropriate 'idealising' constructions. Geometry was no longer characterized as a theory of space in a narrow sense, but as a general theory of *Ordnungssetzungen* (*order posits*) (cf. [Carnap, 1928]).

The philosophical upshot of this evolution was that even in geometry the role of Kantian pure intuition was in decline. Instead of *Anschauung*, general principles of theory construction began to play a central role such as principles of duality and completion. More generally, considerations of practical and theoretical fruitfulness became dominant (cf. [Wilson, 1992; Tappenden, 1995]). This change was grist to the mill of the neo-Kantians. In contrast to orthodox Kantians, the neo-Kantians took the new developments as a confirmation of their Anti-Kantian claim that the Kantian idea of pure intuition had to be abandoned in the light of modern science and mathematics. According to Cassirer, the development of mathematics in the 19th century made a naive intuition-oriented view of geometry untenable. He whole-heartedly welcomed the new developments in geometry.[9]

The methodology of ideal elements was not confined to geometry. Also the domain of algebra underwent a growing variety of idealisational procedures which unfolded the original narrow domain of numbers in a multifaceted way. Maybe the best-known example is the idealisational procedure of *Dedekind cuts* which Cassirer chose as his paradigmatic case.

Let us consider the rational number **Q** as the objects 'antecedently understood', to use an apt terminology of Hempel. As is generally agreed, rational numbers

---

[7]Cf. Wilson's discussion of von Staudt's method of 'concept-objects'.

[8]Cf. [Torretti, 1978, Chapter 2.3].

[9]His discussion of the philosophical relevance of projective geometry in *Substance and Function* may still be considered as one of the best that can be found in the philosophical literature.

are useful concepts, but for certain applications they are less than optimal, for instance for measuring, or for solving polynomial equations. One is in need of more numbers, the so called irrational numbers. In other words, with respect to the problems of solving polynomial equations, the domain of rational numbers does not qualify as an appropriate setting. In order to overcome this unpleasant situation one has to construct new 'ideal' or 'imaginary' numbers that provide solutions for equations for which 'real' solutions do not exist.

In order to construct the missing irrational numbers we may consider the set of 'cuts' of $\mathbf{Q}$, i.e., the set of partitions of $\mathbf{Q}$ in two mutually disjoint and jointly exhaustive subsets such that all elements of the lower cut are strictly smaller than all elements of the upper cut. Obviously, a cut is determined by either its lower or upper cut alone. Hence, we are entitled to carry out the completion of $\mathbf{Q}$ by the lower cuts. Denoting the power set of all subsets of $\mathbf{Q}$ by $P\mathbf{Q}$, a lower cut $q \in P\mathbf{Q}$ may be precisely defined as follows: q is a lower cut iff it satisfies the following two conditions:

1. q is a downset, i.e., for all $a \leq b, b \in q \Rightarrow a \in q$.

2. q has no maximal element.

Denote the set of lower cuts of $\mathbf{Q}$ by $C\mathbf{Q}$. Then there is a canonical embedding of $\mathbf{Q} \xrightarrow{e} C\mathbf{Q}$ by mapping a rational number $r \in \mathbf{Q}$ to the cut $q(r)$ defined by $e(r) := \{a; a < r\}$. This means, rational numbers can be identified with lower cuts whose upper cuts have a minimal element, to wit r itself. Hence the set of rational numbers $\mathbf{Q}$ can be identified with its image $e(\mathbf{Q}) \subseteq C(\mathbf{Q})$. The interesting point is that $C(\mathbf{Q})$ is larger than $e(\mathbf{Q})$—not all lower cuts have the form $e(r)$. [10] A typical example is the cut $\{a; a^2 < 2\}$ which is to be identified with the irrational number $\sqrt{2}$. The set $\{a; a^2 < 2\}$ is not a rational cut since its upper cut $\{a; a^2 \geq 2\}$ has no minimal element. The next step is to identify $C(\mathbf{Q})$ with the set $\mathbf{R}$ of real numbers by showing that the arithmetic of rational numbers $\mathbf{Q}$ can be extended to the arithmetic of $C(\mathbf{Q})$ in such a way that the elements of $C(\mathbf{Q})$ play indeed the arithmetic roles they are designed to play. Summing up we may say that by embedding $\mathbf{Q}$ into $C(\mathbf{Q})$ one has completed the domain $\mathbf{Q}$ of rational numbers by certain 'ideal elements' in such a way that the completed domain $C(\mathbf{Q})$ behaves arithmetically better than $\mathbf{Q}$.

Although for the completed domain of real numbers $\mathbf{R}$ arithmetic works more smoothly than for rational numbers, even $\mathbf{R}$ leaves something to desire, since it is not algebraically complete. In order to get a fully satisfying theory of polynomial equations for which the fundamental theorem of algebra holds, the algebraically incomplete field $\mathbf{R}$ should be replaced by the algebraically complete field $\mathbf{C}$ of

---

[10] Actually, the cardinality of $C(\mathbf{Q})$ is much larger than that of $\mathbf{Q}$: $\mathbf{Q}$ is countable but $C(\mathbf{Q})$ has the cardinality of the continuum.

complex numbers. This may be done by considering complex numbers as pairs (a, b) of real numbers $a, b \in \mathbf{R}$, such that the original real numbers are identified with pairs of type $(a, 0)$. Again, this process may be conceived as an idealising completion process, although of a different kind than Dedekind's. For our purposes these differences are not important, the upshot for a theory of idealisation is that the introduction of new 'ideal' or 'imaginary' elements in geometry or numbers, not only enlarges the domain of objects to be considered, but, and this is the most important characteristics, it enhances the global conceptual features of the domain to be considered. Put in a nutshell, it leads to construction of new, more appropriate settings for doing geometry and algebra. Thus, these theories are not to be confined to their fixed domains of platonic entities, rather they are to be conceived as open fields of idealising constructions. These idealising constructions introduce a wealth of new objects that render untenable any 'intuitionist' conception of mathematics as the theory of an intuitively given unalterable domain of timeless platonic entities. Although it may well be the case that mathematics once started in some intuitive domain, it certainly did not remain confined to it.

One of the deepest philosophical insights of Cassirer's idealist philosophy of science was that constructions such as Dedekind's are not only mathematically interesting technical achievements. Rather, these idealisational constructions are to be considered as the prototypes of idealisational constructions essential for 20th century's mathematics in general.[11] Evidence for this sweeping claim is that 'idealisation' and 'completion' in the sense of finding 'appropriate settings' for the problems one is studying is now a routine part of the mathematicians's work. Typical are the following remarks of the mathematician Horst Herrlich;[12] After having listed nine assertions concerning set-theoretical topology, he remarks:

> Although we would like the above statements to be true, we know that none of them is [true]... in the category *Top* of topological spaces and continuous maps. However, there exist settings—more appropriate it would seem—in which the above statements are valid. The category *Top* can be decently embedded in larger, more convenient categories such that ... the above statements are not only true but, in fact, special cases of more general theorems. [Herrlich, 1976, p. 265]

---

[11] For instance, the proof of one of the most famous theorems of 20th century mathematics, Stone's representation theorem, may be considered as a generalisation of Dedekind's cut construction.

[12] I'd like to emphasise that Herrlich's remarks have no 'philosophical' intentions. His article may be characterized as a mixture of a survey and a research paper. Philosophers of mathematics certainly do not belong to its intended audience. The general tenor of Herrlich's paper is in no way original, analogous remarks for other research areas may easily be found in many places.

Then he goes on and proposes other categories[13] than *Top* as more appropriate settings for doing topology. These categories are more appropriate settings for doing topology since in them the above mentioned 'assertions concerning set-theoretical topology' become provable theorems. Thus, category theory as a general theory of local mathematical frameworks offers a framework in which mathematicians can discuss problems of appropriate setting in a way that goes beyond a subjectivist presentation of personal whims and preferences. Of course, there is still room for negotiation. It may turn out that there are good reasons for wishing one of two incompatible theorems to be true. An interesting case is the axiom of choice (AC) and the axiom of determination (AD): for many domains of mathematics the axiom of choice (AC) seems to be virtually indispensable (e.g. topology) while for others, the axiom of determination (AD), which may be considered as the opposite to (AC), appears more attractive. Examples like these can be easily multiplied. They show that mathematics of 20th century has been fully aware of the importance of the method of ideal elements, or, more generally, of idealisation. Unfortunately, this issue has yet to find the attention it deserves from the side of philosophy of mathematics:

> The official position, dominant since the start of this (the 20th) century, maintains that any self-consistent domain is equally worthy of mathematical investigation; preference for a given domain is justified only by aesthetic considerations, personal whim or its potential physical applications. [Wilson, 1992, p. 152]

Although the 'official position' meanwhile may have lost some of its strength, having given room for some less global, more local and specific considerations, it is remarkable how far ahead Cassirer's account was compared with that of most philosophers of mathematics who still stick to the 'official position'.[14] His account may be characterized as a kind of a general pragmatics of idealisation and completion.

As a last mathematical evidence for the importance of idealising completions in mathematics I'd like to mention Stone's representation theorem that provided the first non-trivial relation between topology and logic.

Let $\mathbf{B}$ be any Boolean algebra. Every $b \in \mathbf{B}$ defines an ideal, to wit, the set $r(b) := \{a; a \leq b, b \in \mathbf{B}\}$. Denoting the set of ideals of $\mathbf{B}$ by IDEAL($\mathbf{B}$) one ob-

---

[13] A category may be informally characterised as a local universe of mathematical discourse, for a detailed discussion of categories and their role in mathematics see[Adámek *et al.*,].

[14] It is a pity that in the recent work on the issue of 'appropriate settings' for mathematical theories Cassirer's contributions are completely ignored. Wilson [1992] asserts in a footnote that it is worth mentioning that various philosophers attempted to extend the 'principle of continuity' (i.e., the introduction of ideal elements) into general philosophy of language, e.g. Cassirer in [*Substance and Function*]. This is a somewhat strange remark: *Substance and Function* deals with many things, but certainly not with a 'general philosophy of language'.

tains a map $r : \mathbf{B} \longrightarrow \mathrm{IDEAL}(\mathbf{B})$. One can show that in general $\mathrm{IDEAL}(\mathbf{B})$ is larger than $\mathbf{B}$, since there are ideals that are not of the form $r(b)$. These 'non-principal' ideals correspond to the non-rational numbers in the analogous Dedekind completion of rational numbers. In other words, $\mathrm{IDEAL}(\mathbf{B})$ may be conceived as a completion of $\mathbf{B}$. As was shown by Stone in the thirties, this completion defines a bridge from the theory of Boolean algebras to the theory of totally disconnected Boolean (or Stone) topological spaces in the sense that $\mathrm{IDEAL}(\mathbf{B})$ uniquely defines a topological space whose lattice of open and closed sets is isomorphic to $\mathbf{B}$ (cf. [Davey and Priestley, 1990]). Thereby one obtains a 'concrete' set-theoretical representation of the 'abstract' Boolean algebra B.

Stone's theorem has been considered as one of the most important theorems of 20th century's mathematics (cf. [Johnstone, 1982]). This is not the place to discuss this assessment in any detail but it certainly evidences that idealising completions are of utmost importance for contemporary mathematics.

Guided by ST Cassirer did not confine the method of ideal elements to mathematics. According to him, the same kind of idealising completions can be found in the empirical sciences. This will be discussed in the next section.

4   Idealisation in Mathematical Physics

Critical idealism argued for the importance of idealisations in physics and other empirical sciences from an empirical perspective, so to speak. It is a fact that the advanced sciences make heavy use of mathematics, and it is the task of philosophy of science to make sense of this fact. For instrumentalist and empiricist currents of philosophy of science the employment of advanced mathematics in all areas of science presents a conceptual difficulty since according to them scientific concepts have only the task of reproducing the given facts of perception in abbreviated form [?, p. 148]. If this were really the case, the task of philosophy of physics would be achieved, if every concept of a physical theory had been dissolved into a sum of perceptions such that this sum could be used to recover the full realm of empirical facts falling under that concept (cf. [?, p. 151]). But such a replacement of mathematical concepts by perceptual or observational ones is virtually impossible:

> The theories of physics gain their definiteness from the mathematical form in which they are expressed. The function of numbering and measuring is indispensable even in order to produce the raw material of 'facts' that are to be reproduced and unified in theory.
>
> ...
>
> [For] it is precisely the complex mathematical concepts, such as possess no possibility of direct sensuous realisation, that are continually used in the constructions of mechanics and physics. Conceptions,

which are completely alien to intuition in their origin and logical properties, and transcend it in principle, lead to fruitful applications within
intuition itself. This relation finds its most pregnant expression in the
analysis of the infinite, yet is not limited to the latter. [?, p. 116]

According to Cassirer, it is this complex intertwinement of the 'factual' and
'theoretical' on which the edifice of physics is based (cf. [Cassirer, 1953, p.130]).
This claim should not be dismissed as the fancyful inventionof a speculative philosopher. For instance, the renowned physicist Henry Margenau [15] maintained a similar
conception distinguishing perceptibel matters of facts and theoretical (or symbolic)
constructs between which a kind of dialectical movement takes place.

> Physical inquiry in its essential form moves along a peculiar cycle: it
> starts with definite, perceptiable matters of facts; proceeds from there
> into a field in which at least some of the elements of operations are
> not directly perceptible, where there is greater freedom from empirical
> constraints; and finally it ermerges again in the realm of perceptible
> facts. [Margenau, 1935, p. 57]

According to Marganau, the distinction between 'perceptible facts' and 'theoretical constructs' cannot be drawn in the way that the former are characterised as
'real' while the latter are said to be 'mere fictions'. They are intertwined in a
complex manner that renders obsolete this traditional distinction (cf. [?, p. 164]).
Already in *Substance and Function*, Cassirer had pointed out that motion in the
sense of physics cannot be conceived as a purely sensuous phenomenon that can
be perceived as such. It demands the continuous and homogeneous space of pure
geometry as a foundation. As long as space is conceived just as a sum of visual or
tactile impressions it does not allow 'motion' in the sense of physics (cf. [Cassirer,
1953, p. 118]). In brief, motion is a fact of conception, not of perception. However, it is important to note that the idealising method of empirical science should
not simply be conceived as a replacement of the directly observable experiences
by their ideal limit cases. This would sugggset that the objects empirical science is
dealing with are in line with the objects of perception. Thereby idealisation would
boil down to not much more than approximation. Idealisation could be characterised as a continuation of empirical observatio. Cassirer emphatically insists that
this is not the case. The ideal elements to be introduced ar enot just some other
things we add to the 'real' things. Rather, they express a certain way we deal with
the 'real' things. This may become plausible, if we consider in some detail the
construction of points as limiting elements of other more basic 'empirical' objects
called regions.

---

[15] In later years, Margenau was to become a colleague of Cassirer's at Yale. He was one of the few
working physicists of the 20th century who took seriously neo-Kantian philosophy of science, see his
[?].

The underlying problem may be described as follows: In physical geometry and other sciences the talk of points as basic elements is ubiquitous. Nevertheless, in 'reality' one never meets points. They are idealisational constructs. The measured values of a physical magnitude never assume points as their values simply because this would amount to absolute precision which real science never reaches. Instead, in real science, the measured values are assumed to be located in some more or less extended intervals. Usually, these intervals are considered as sets of points. Thus, even if one admits that our measuring methods never reach the points, we are accustomed to consider them as the basic building blocks of space, time, space-time, and other generalised spatial structures used in science. From a strictly empirical point of view, however, points appear to be rather contrived entities. It would be too simple, however, to consider them just as convenient mathematical fictions that 'somehow' play the role they are assumed to play. This would amount to a strict separation between the domain of empirical reality on the one hand and the domain of mathematics on the other hand whereby it becomes impossible to bring them together again by somehow establishing a link between them by stipulation. The question is how to avoid the standard dichomotic account.

An answer, or at least some hints, where one may look for an answer, can be found in Whitehead's method of 'extensive abstraction' [Whitehead, 1929]. Whitehead was probably the first who attempted to replace points as fundamental entities of spatial and temporal structures by objects that were empirically more accessible. In the case of space and spacetime these entities may be characterized as *regions*. Intuitively, a spatial region may be described as a more or less well-shaped part of space. Whitehead's programme was to take regions instead of points as the basic fundamental building blocks and to construct points and their geometric relations from regions and their relations. Whitehead only gave an informal sketch of how this might work but his account can be reconstructed in a formally rigorous way (cf. [Mormann, 1998]).

The interesting point is that this construction of points from regions can be conceived as a generalisation of Dedekind's method of cuts. That is to say, the insertion of points as ideal or limit elements of the realm of regions is an idealising completion process analogous to the construction of real numbers from rational ones. Ignoring the technical details it goes like this: Assume $\mathbf{W}$ to be a complete Boolean algebra of regions: This means that for two regions a and b a relation $\leq$ is defined such that $(\mathbf{W}, \leq)$ satisfies the axioms of a complete Boolean algebra. The relation '$a \leq b$' is to be read as 'The region a is part of the region b'. One observes that the existence of points is not presupposed. In order to introduce points, one needs a further relation $\ll$ such that $a \ll b$ is to be intuitively read as 'a is an *interior* part of b' in the sense that a does not touch the boundary of b. [16]

---

[16] Actually it is sufficient to take the relation $\ll$ of interior parthood as the only primitive relation, since the standard parthood relation $\leq$ can be defined in terms of $\ll$ (cf. [Mormann, 1998]).

A set q of regions is a 'round ideal' iff it is a downset of regions, and for $a \in q$ there is always a region b such that $a \ll b \in q$. The set of round ideals $\text{IDEAL}_r(\mathbf{W})$ of $\mathbf{W}$ corresponds to the Dedekind $C(\mathbf{Q})$ of $\mathbf{Q}$, while $\mathbf{W}$ corresponds to $\mathbf{Q}$. An embedding $W\text{–e}\rightarrow \text{IDEAL}_r(\mathbf{W})$ is given by $b \longrightarrow \{a; a \ll b\}$. $\text{IDEAL}_r(\mathbf{W})$ can be topologically represented in the sense that there is a topological space[17] $(pt(\text{IDEAL}_r(\mathbf{W})), \mathbf{O}(\mathbf{pt}(\text{IDEAL}_\mathbf{r}(\mathbf{W}))))$ such that a region $a \in \mathbf{W}$ is represented by a regular open subset $e(a) \in O(pt(\text{IDEAL}_r(\mathbf{W}))) = \text{IDEAL}_\mathbf{r}(\mathbf{W})$. Although this construction is technically more complicated than that of Dedekind cuts, it follows the same pattern. The upshot is that by this idealising completion regions can be represented by point sets. This is a necessary condition if one wants to speak of 'motion' in a scientific sense. More precisely, the completion of $\mathbf{W}$ by $\text{IDEAL}_r(\mathbf{W})$ allows to conceive processes as continuous mappings $I \longrightarrow X$ of a time interval I into a topological space X defined by $\text{IDEAL}_r(\mathbf{W})$.

Before leaving the general discussion of idealising completions it may be expedient to emphasise once more the formal similarity of the three examples considered so far. All can be described as embeddings $X\text{–e}\rightarrow I(X)$ of the domain X into an ideal completion $I(X)$:

1. The completion of $\mathbf{Q}$ to $\mathbf{R}$ by a map $\mathbf{Q} \longrightarrow \mathbf{R}$.

2. The completion of $\mathbf{B}$ to $\text{IDEAL}(\mathbf{B})$ by a map $\mathbf{B} \longrightarrow \text{IDEAL}(\mathbf{B})$.

3. The completion of $\mathbf{W}$ to $\text{IDEAL}_r(\mathbf{W})$ by a map $\mathbf{W} \longrightarrow \text{IDEAL}_r(\mathbf{W})$.

Of course, not just every map will $X \longrightarrow I(X)$ will do as an honest completion, certain structural requirements have to be satisfied.[18] Thus one may speak of a general theory of completions in which the examples (1)–(3) can be considered as paradigmatic cases.[19] This corroborates Cassirer's claim that Dedekind-like completion methods are among the important foundational syntheses on which mathematics *and* empirical sciences are grounded. This corroboration is the more compelling the more one delves into the intricacies of modern mathematics and science. Without going into the details of this general theory of completion methods, one may note that idealising completions can be conceived of as *representations*: the incomplete and 'non-ideal' manifold X is *represented* by the completed 'ideal' manifold $I(X)$. For instance, an incomplete perceptual manifold

---

[17] A topological space $(X, O(X))$ is a set X endowed with a set $O(X)$ of open subsets of X satisfying certain properties. Details can be found in any textbook on topology. For a succinct presentation, see [Davey and Priestley, 1990]. If the topology is understood, a topological space $(X, O(X))$ is denoted by X.

[18] A comprehensive general discussion of the many kinds of completions occurring in various areas of mathematics may be found in (cf. [Adámek *et al.*, , III.12]).

[19] It is a nice terminological coincidence that in these examples idealisations go together with ideals. This is not always the case. For instance, in topological contexts, idealisations are often described as compactifications, which have not direct idealising connotations.

of experiences is represented by a completed conceptual manifold. This kind of construction is ubiquitous in physics and other empirical sciences.

The Whiteheadean construction of physical space is a typical example but in no way the only one. Quite the contrary, physical space is only one among a wealth of spatial constructions employed in the empirical sciences. One may even contend that it does not provide the best example for this kind of constructions, since its overly simple appearance may hide its conceptually constructed character. This misapprehension is avoided when one considers the construction of state spaces of empirical theories in general. State spaces are defined with respect to certain systems, i.e., a state space is always a state space of a system or a kind of systems. The concept of a *system* is taken as primitive. Examples of systems are provided by mechanical or thermodynamic systems such as a particles, projectiles, pendulums, planets, gases, liquids, lasers. Even entities as large as galaxies may be considered as systems, or the universe itself taken as the largest possible system. Generally, a system S is an appropriately chosen chunk of the world taken to be the object of theoretical investigation. Systems are assumed to be possibly in different states. For instance, an atom considered as a system in the sense of quantum theory may be in an excited state or not. In order to be accessible to theoretical considerations at all, a class of *possible states* of S must be selected. This class of possible states is denoted by $\Sigma(S)$ or simply Σ if S is understood. $\Sigma(S)$ is called the state space of S.

Here, 'possible" is to be understood in the weak sense of logical possibility. That is to say, some of the elements of Σ may well turn out to be really impossible, i.e., physically impossible states for S. For instance, in first approximation, the state space $\Sigma(o)$ of a material object o may be taken to be the whole universe, even if most places in the universe are physically inaccessible for o. The state space of a system S serves only as general stage on which its story is rehearsed. It is not assumed that S has to occupy all possible locations during the play. Quite the contrary. It is a crucial task of the theory to select certain areas as the 'really possible" states and to classify their complement as a sort of no-go area for S. Thereby, a modal component is introduced in the theory's framework. For a rough and preliminary distinction between possible and impossible states purely set theoretical methods will suffice, but for a more refined determination, in particular for the distinction between possible and impossible *processes*, more refined (geometrical) structures of the state space come into play. Let us consider some examples. The first is Cassirer's and once again rehearses the case of 'physical space':

> The individual positions of Mars, which Kepler took as a basis ... do not in themselves alone contain the thought of the orbit of Mars; and all the heaping up of particular positions could not lead to this thought, if there were not active from the beginning ideal presuppositions though which the gaps of actual perception are supplemented.

What sensation offers is and remains a plurality of luminous points in the heavens; it is only the pure mathematical concept of the ellipses, which has to have been previously conceived, which transforms this discrete aggregate into a continuous system. Every assertion concerning the unitary path of a moving body involves the assumption of an infinity of possible places; however, the infinite obviously cannot be perceived as such, but first arises in intellectual synthesis and in the anticipation of a universal law. Motion is gained as a scientific fact only after we produce by this law a determination that includes the totality of the space and time points, which can be constructively generated, in so far as this determination coordinates to every moment of continuous time one and only one position of the body in space. [Cassirer, 1953, p. 118–119]

If even for physical space—conceived as a state space—'real' and 'not-real' elements are inseparately interwoven, this holds *a fortiori* for general state spaces [Cassirer, 1953, p. 117]. Physical space is only a first evidence that idealising elements always play an essential role. For instance, consider an elementary thermodynamical system S characterized by the two quantities of volume and pressure only. As a first approximation of the state space $\Sigma(S)$ one may take a 2-dimensional Euclidean plan **E** having an orthogonal base consisting of the two vectors V (volume) and P (pressure). Since negative volume and pressure do not make sense the 'really possible' states of S are to be found only in the first quadrant of **E**. Actually, further constraints will play a role. If we assume the ideal gas law to hold, the product $S(V) \cdot S(P)$ must be constant for all 'really" possible states of S. Hence, the state space $\Sigma(S)$ of possible states of S is the hyperbola defined by the equation $S(V) \cdot S(P) = constans$.

State spaces are not empirical objects given by nature. Rather they depend on the theories used. Depending on the theory different state spaces for the 'same' S may be obtained.[20] In any case, the first step for theoretically understanding the behaviour of any empirical system S consists in providing an appropriate state space. In other words, a system S enters the theoretical realm only if it S represented by an appropriate state space. Now, as is already suggested by the term *space*, $\Sigma(S, T)$ is usually not simply a set but a space, i.e., a set endowed with some geometric structure. This structure is employed to differentiate between really possible and really impossible states of the system.

What is really interesting, however, is not the state a system occupies or does not occupy, but rather the processes a system may run through. State spaces provide useful representations of empirical processes by representing them as paths in the state space. Mathematically a path is defined as a map $f : I \longrightarrow \Sigma(S, T)$

---

[20]Hence, one might have written $\Sigma(S, T)$ instead of $\Sigma(S)$.

of the unit interval I into the state space $\Sigma(S, T)$. Only a few paths represent processes allowed according to the laws of the theory, most paths represent forbidden ones, i.e., those that are impossible according to the theory. It may be considered as the essential task of the theory to distinguish the possible and the impossible ones. In order to do this the geometric structure of the representing state space has to be brought into play. Typically, constraints on admissible paths are defined by geometric concepts such as vector fields, differential and tensor forms. All these devices are constructions based on the introduction of ideal elements. The continuous motion of a body in physical space is just the most elementary case, but I hope to have shown that it already suffices to grasp the philosophical essence of the huge variety of spatial constructions used in science.

Up to now we have discussed only conceptual idealisations, i.e., idealisations that amount to an embedding or completion of an empirical manifold of perceptions or experiences into a conceptual framework that is richer and better structured than the empirical manifold we started with. This might suggest that idealisation in the sense of Cassirer is merely a conceptual activity 'in the head'. I think, this would amount to a misunderstanding or at least a too narrow interpretation succumbing to the old anti-idealist prejudice according to which idealism is always concerned with the 'mental' (cf. [Haack, 2002]). Instead, I'd like to propose to understand idealising as an activity to construct 'appropriate settings' (cf. [Wilson, 1992]) for certain scientific purposes. According to Cassirer, it is a matter of science what counts as appropriate settings and how they are constructed, and what are the means for their constructions. It may well be the case that science will invent hitherto unknown methods of idealising, i.e., constructing 'appropriate settings' for its purposes. Examples are the various methods of simulation that recently have been developed and play an ever more important in many areas of science.[21] There is no reason to assume that for Cassirer's philosophy of science the arsenal of idealising methods is a priori restricted to purely conceptual idealisations. Rather, idealising may involve machines, simulations and possible other devices that bring about appropriate settings for interesting stable phenomena. Thus we may speak, somewhat paradoxically, of material idealisations. I think this is not a too far-fetched generalisation of Cassirer's account. For him, the most important feature of scientific conceptualisation has always been the serial character of concepts through which they lawfully connect a manifold of isolated experiences [?, p. 148]. From an algorithmic perspective, a serial concept is noth-

---

[21] Another option is to conceive the so called 'nomological machines' of Cartwright as idealising devices [Cartwright, 1995]. Indeed, as Pickering showed sometime ago, there are far-reaching similarities between the building of highly complicated nomological machines like bubble chambers of particle physics and the construction of conceptual systems such as Hamilton's quaternions (cf.[Pickering, 1996]). On a more elementary level, one could read Lakatos's discussion of establishing a context for a valid proof for Euler's theorem as a series of attempts to build a smoothly running conceptual machine (cf. [?]).

ing but a conceptual machine that produces for a given input an output following its internal rules. According to *ST* this holds for scientific concepts in general, they may be empirical or mathematical ones. More precisely, physical concepts only continue what is evident for mathematical concepts:

> ... the physical concepts only carry forward the process that is begun in the mathematical concepts, and which here gains full clarity. The meaning of the mathematical concept cannot be comprehended, as long as we seek any sort of presentational correlate for it in the given; the meaning only appears when we recognize the concept as the expression of a *pure relation*, upon which rests the unity and continuous connection of the members of a manifold. The function of the physical concept also is first evident in this interpretation. The more it disclaims every independent perceptible content and everything pictorial, the more clearly its logical and systematic function is shown. [Cassirer, 1953, p. 166]

Thus, the sameness thesis *ST* may be conceived as providing an appropriate epistemological perspective for dissolving the problem of the applicability of mathematics in empirical science. After all, the applicability of mathematics appears as a miracle only if the spheres of mathematics and the real world are totally separated. This separation, however, is nothing but the result of a bad metaphysics that erroneously reifies methodological differences. From the perspective of *ST*, the concepts used in both area are based on the same foundational syntheses. More specifically, conceptualisation in mathematics as well as in empirical science amounts to idealising constructions of appropriate settings (completed idealized manifolds) for which lawful regularities and stable phenomena obtain.

5   Concluding Remarks

For a general assessment of Cassirer's 'critical idealist' philosophy of science it is expedient not only to discuss some of its more specific technical theses such as *ST* but also to offer an attempt to locate critical idealism on the general map of 20th century's philosophy of science. In this respect I'd propose to conceive critical idealism as a kind of moderate conventionalism. This is suggested by Cassirer's contention that ideal constructions may be characterized as conventions. From the critical idealist's perspective, conventionalism is the thesis

> that thought does not proceed merely receptively and imitatively, but develops a characteristic and original spontaneity. This spontaneity is not unlimited and unrestrained; it is connected, although not with the individual perception, with the system of perceptions in their order and connection. [Cassirer, 1953, p. 187]

Occasionally, Cassirer even adopts an explicitly instrumentalist stance:

> The objects of physics: matter and force, atom and ether can no longer
> be misunderstood as so many new realities for investigation, and real-
> ities whose inner essence is to be penetrated, when once they are rec-
> ognized as instruments produced by thought for the purpose of com-
> prehending the confusion of phenomena as an ordered and measurable
> whole. [Cassirer, 1953, p. 166]

An important feature of the critical idealist's modest conventionalism is its relational holistic component: although the manifolds of our experiences are atomic in
the sense that sensuous experiences can be isolated from each other, the idealised
completed manifolds of our conceptualisations are connected in the sense that a
theory arising from such an idealised manifold is not confronted with isolated experiences but with a relationally conceived reality as a whole.

Locating neo-Kantian philosophy of science in the neighbourhood of conventionalism is not intended to offer a full description of this philosophical current
but at least it is a starting point for overcoming the allegedly unbridgeable abyss
between idealist and analytic philosophy of science, which has hindered a fruitful
discussion between both currents for such a long time.

BIBLIOGRAPHY

[Adámek et al.,] J. Adámek, H. Herrlich and G. Strecker. *Abstract and Concrete Categories*. New York, John Wiley., 1990.

[Carnap, 1928] R. Carnap. *The Logical Structure of the World (Aufbau)*. Berkeley, University of California Press, 1928 (1967).

[Cartwright, 1995] N. Cartwright. Ceteris Paribus Laws and Socio-economic Machines. *The Monist* **78**, 276–294, 1995.

[Cassirer, 1907] E. Cassirer. Kant und die moderne Mathematik. *Kant-Studien*, **XII**, 1–49, 1907.

[Cassirer, 1953] E. Cassirer. *Substance and Function & Einstein's Theory of Relativity*. New York, Dover, 1910; 1953.

[Cassirer, 1929] E. Cassirer. *Die Philosophie der Symbolischen Formen, Band 3*. Darmstadt, Wissenschaftliche Buchgesellschaft, 1929 (1980).

[Cassirer, 1937] E. Cassirer. *Ziele und Wege der Wirklichkeitserkenntnis, Nachgelassene Manuskripte und Texte Band 2*. Hamburg, Meiner Verlag 1937 (1999).

[Corfield, 2003] C. Corfield. *Towards a Philosophy of Real Mathematics*. Cambridge, Cambridge University Press, 2003.

[Davey and Priestley, 1990] B.A. Davey and H.A. Priestley. *Introduction to Lattices and Order*. Cambridge, Cambridge University Press, 1990.

[Friedman, 1999] M. Friedman. *Logical Positivism Reconsidered*. Cambridge, Cambridge University Press, 1999.

[Haack, 2002] S. Haack. Realisms and Their Rivals: Recovering Our Innocence, *Facta Philosophica*, **4**(1), 67–88, 2002.

[Herrlich, 1976] H. Herrlich. Some Topological Theorems which Fail to Be True. In E. Binz, and H. Herrlich (eds.), *Categorical Topology*, pp. 265–285. Volume 540 of Lecture Notes in Mathematics, Springer-Verlag, 1976.

[Johnstone, 1982] P. T. Johnstone. *Stone Spaces*. Cambridge University Press, Cambridge, 1982.

[Lakatos, 1976] I. Lakatos. *Proofs and Refutations. The Logic of Mathematical Discovery*, Imre Lakatos, Elie Zahar, and John Worrall, eds. Cambridge, Cambridge University Press, 1976.
[Margenau, 1935] H. Margenau. Methodology of Modern Physics, *Philosophy of Science*, **2**, 164–187, and 48–72, 1935.
[Margenau, 1950] H. Margenau. *The Nature of Physical Reality*. A Philosophy of Modern Physics, New York, McGraw-Hill, 1950.
[Mormann, 1997] T. Mormann. Der Aufbau der wissenschaftlichen Wirklichkeit bei Cassirer. *Logos, Zeitschrift für systematische Philosophie*, **5**, 268–293, 1997.
[Mormann, 1998] T. Mormann. Continuous Lattices and Whiteheadian Theory of Space, *Logic and Logical Philosophy*, **6**, 35–54, 1998.
[Mormann, 1999] T. Mormann. Idealistisische Häresien in der Wissenschaftsphilosophie: Cassirer, Carnap, Kuhn. *Journal of General Philosophy of Science*, **30**(2), 233–270, 1999.
[Nagel, 1971] E. Nagel. The Formation of Modern Conception of Formal Logic in the Development of Geometry. In *Teleology Revisited*, New York, Columbia University Press, 1971.
[Nowakowa and Nowak, 2000] I. Nowakowa and L. Nowak. *The Richness of Idealization*, Poznan Studies in the Philosophy of the Sciences and the Humanities: Amsterdam, Rodopi, 2000.
[Pickering, 1996] A. Pickering. *The Mangle of Practice*. Chicago, Chicago University Press, 1996.
[Richardson, 1997] A. Richardson. *Carnap's Logical Construction of the World*, Oxford, Oxford University Press, 1997.
[Russell, 1903] B. Russell. *The Principles of Mathematics*. London, Routledge and Kegan Paul, 1903.
[Scheffler, 1967] I. Scheffler. *Science and Subjectivity*. Indianapolis, Bobbs-Merrill, 1967.
[Smith, 1959] D.E. Smith. *A Source Book in Mathematics*. New York, Dover, 1959.
[Tappenden, 1995] J. Tappenden. Extending Knowledge and 'Fruitful Concepts": Fregean Themes in the Foundations of Matehmatics. *Nous*, **29**, 427–467, 1995.
[Torretti, 1978] R. Torretti. *Philosophy of Geometry from Riemann to Poincaré*. Reidel, Dordrecht, 1978.
[Wilson, 1992] M. Wilson. Frege: The Royal Road from Geometry. *Nous*, **26**, 149–80, 1992.
[Wright, 1992] C. Wright. *Truth and Objectivity*. Oxford, Blackwell, 1992.
[Whitehead, 1929] A.N. Whitehead. *Process and Reality*. London, Macmillan, 1929.

Thomas Mormann

Department of Logic and Philosophy of Science
University of the Basque Country UPV/EHU
P.O. Box 1249
20080 Donostia-San Sebastián, Spain
Email: ylxmomot@sf.ehu.es

Cassirer's Critical Idealism
A Comment on Thomas Mormann

MAARTEN VAN DYCK AND ERIK WEBER

1 Introduction

Let us begin with stating that we are in agreement with the overall tendency of Mormann's paper. We will give some complimentary remarks on Cassirer's philosophical programme, and on the place therein of what Mormann calls the sameness thesis, focusing more than Mormann does on the neo-Kantian aspect of Cassirer's position (Section 2). We hope that these necessarily preliminary discussions will help to show the ingenuity and continuing value of Cassirer's work, and concurrently of 'idealist' philosophy of science.[1] To this end it will also be fruitful to draw attention to some aspects of Cassirer's views that are not mentioned by Mormann, especially since in them Cassirer proves to be a precursor of some tenets of the nowadays popular model-theoretic approach to scientific theories (Section 3). We will also comment on Mormann's characterization of Cassirer's critical idealism as a moderate conventionalism by comparing Cassirer with Henri Poincaré (Section 4).

One of Mormann's underlying aims is to argue that the work of Cassirer has been unduly neglected. We wholeheartedly agree. Let us add a telling example, which helps to set the stage for our discussion of Cassirer's program. In his *A Hundred Years of Philosophy* (first published in 1956) John Passmore spent a mere six paragraphs on Cassirer, in a chapter entitled *'recalcitrant metaphysicians'*. These paragraphs end with a blunt denial of Cassirer's status as "a 'philosopher', as that word is now commonly understood by British philosophers", [Passmore, 1966, p. 318]. It is hard to understand on what grounds, other than ignorance, a thinker that engaged so closely with the philosophy of mathematics of Bertrand Russell—*the* arch-British philosopher—as Cassirer does in *Kant und die moderne Mathematik* (1907) and *Substance and Function* (1910/1953), could be denied the status of

---

[1] There has been not too much critical discussion of Cassirer. Krois [1987] and Friedman [2000] are very welcome exceptions to which we refer the reader interested in more thorough discussions of Cassirer's philosophy. We will limit ourselves to Cassirer's position up to 1910—Krois [1987] argues forcefully that after that date Cassirer's thought went through a genuine transformation, resulting in the Philosophy of Symbolic Forms. Nevertheless, also in his later years Cassirer retained many of the insights expounded in Substance and Function, but rather placed them in a different overall orientation.

Laws and Models in Science, 161–171.

'a philosopher as understood by British philosophers'. His inclusion in the chapter 'recalcitrant metaphysicians' is justified by his avowed idealism, but as Mormann already warns in his paper, such a denomination must be approached with care—as we will point out, Cassirer's project is much better characterized as anti-metaphysical than 'recalcitrant'.

2   Cassirer's Critical Idealism and the Sameness Thesis

Cassirer states his sameness thesis only at the end of his article on 'Kant and modern mathematics' (KUMM), rather as a programmatic statement than as a conclusion. As the quotation given in Mormann's paper makes clear, it is intended as a kind of summary of his critical idealism. 'Substance and Function' (SF) takes up this thesis and gives a much more detailed exposition of its content. Nevertheless, before dealing with the precise content of the thesis, it is useful to go back to KUMM, as it is proposed there in the course of a discussion on the right way of interpreting Kant's critical philosophy; a discussion that provides an interesting background to the question how to interpret Cassirer's critical philosophy.

2.1

The main body of KUMM consists in a detailed exposition of recent work in the foundations of mathematics and its relation to the logic of relations. The main conclusion is that intuition has to be denied an essential foundational role in all areas of mathematics. Still, Cassirer wants to maintain that this does not invalidate the main tendencies of Kantian philosophy since these same foundational studies still show the essential role played by pure reason (under the guise of logic of relations) in constituting the necessary conditions for the possibility of mathematics. As is clear by the disappearance of the pure forms of sensible intuition, a considerable amount of reinterpretation is necessary to hold up Kantian philosophy in face of these mathematical developments. Since the role of intuition is taken over by logic, the suspicion naturally arises that we are only left with analytic a priori judgements and synthetic a posteriori judgements, with no room for synthetic a priori judgements. Cassirer forcefully disagrees: the logical nature of mathematical judgements in no way detracts from their synthetic character if we understand (with Kant) by synthesis the act of uniting distinct impressions (*Vorstellungen*) and comprehending their variety in one act of understanding (KUMM, 36). The logical principles of serial order which underlie all mathematics qualify as synthetic since they draw together different elements under the determination of one series (cf. especially chapters I,II,III of SF). This synthetic character is of the utmost importance for safeguarding the most important achievement of Kant's philosophical system: rendering intelligible the possibility of *applied* mathematics in a physical science of nature, and thus delineating the field of objective knowledge.

Formal logic is clearly insufficient for dealing with the problem of applied mathematics, since it is only concerned with formal validity, not with objective knowledge (*gegenständlichen Erkenntnis*). The research on this problem, which is the area of critical philosophy, is directed towards the foundation of a *transcendental logic*, which seeks to establish the conditions of the possibility of such objective knowledge by laying bare the logical structures present in all experience (i.e. the synthetic a priori judgements). Having argued that all mathematical knowledge is made possible by serial principles, Cassirer now announces in KUMM that the same foundational synthesises underlie the construction of experimental knowledge (sameness thesis), thus pointing towards an explanation of how mathematics can be applied to nature.

2.2

SF falls apart in two parts, of which the first is devoted to the problem of concept formation (*Begriffsbildung*). In it Cassirer tries to substantiate his sameness thesis first formulated in KUMM. To this end he gives detailed expositions of how mathematical and scientific concepts are constructed, with one constantly recurring theme: the constitutional role of relational structures. Genuine definitions for concepts are not reached by giving a list of properties but by pointing to the generating relations responsible for their true meaning.

The concept of irrational numbers, as explained by Mormann, is a case in point. After introducing the procedure of Dedekind cuts, Cassirer remarks that irrational numbers have 'no other function and meaning than to represent conceptually this determinateness of division [by the cuts] itself. The new number, in this form of derivation, is thus not arbitrarily conceived, nor is it introduced as a mere 'symbol'; but *it appears as the expression of a whole complex of relations*, which were first deduced with strict logic' (SF, 59—our emphasis). What is important is that there is a conceptual rule which can be seen to generate these new 'ideal elements'. Mormann's paper makes admirably clear that such an introduction brings with it important advantages by providing mathematics with a more 'appropriate setting'. This is indeed very important for Cassirer, since these ideal elements are only introduced because they render a given domain conceptually more determinate.

The same kind of constructive generation can be seen to underlie all physical concepts. As again explained by Mormann, the concept of motion provides a good example.

> Motion is gained as a scientific fact only after we produce by this law
> a determination that includes the totality of the space and time points,
> *which can be constructively generated*, in so far as this determination
> coordinates to every moment of continuous time one and only one
> position of the body in space (SF, 119—our emphasis).

But SF contains a wealth of other examples. Cassirer discusses the concepts of atoms, inertial mass, space and time, energy, and chemical valency as so many instances of the introduction of ideal elements by means of serial principles. Modern high energy physics could offer many other convincing examples. All these concepts enable the formulation of laws, and it is here that they find their ultimate justification—as Mormann would put it, they provide appropriate settings for describing nature.

As both mathematical and physical concepts rest ultimately on the introduction of ideal elements by similar means, a beginning can be made of understanding the 'transference of structures, whose whole content is rooted in a connection of pure ideal constructions, to the sphere of concrete factual being' (SF, 117). Since physical concepts appear to be ways of transforming the 'given' so that it can be assigned a fixed place within several series, mathematics, which is precisely the study of such serial forms, should be applicable to physical descriptions.[2] Of course, this can only be half of the story—although Mormann does not mention this—since it still has to be explained why nature should be describable by physical concepts... It would lead us too far to go into Cassirer's solution to this problem, which is to be found in the second part of SF.[3] We will only mention that having dropped the pure forms of sensible intuition, Cassirer cannot transfer Kant's analysis without profound modifications that create their own problems.[4]

2.3

Let us now take up the issue of Cassirer's idealism. As explained Cassirer locates his project in the Kantian critical framework of discovering synthetic a priori elements in experience. It is important to stress that for him the prior aspect has to do with synthetic *acts*, i.e. the bringing together of different elements under one determination. This gives rise to an essentially dynamic view, in which the precise nature of the prior elements of experience varies, and in which only their general form remains, i.e. the postulation of unity in nature.[5] 'Order is not something which can be immediately pointed out in sense-impressions but is rather something which belongs to them only by virtue of intellectual relations' (SF, 44).

As Mormann already stresses, Cassirer's position must not be equated with a naïve idealism that makes everything mind-dependent. His idealism 'does not mean physical dependency on particular thinking subjects, but logical dependency on the content of certain universal principles of all knowledge' (SF, 298). Tellingly, Cassirer often refers to his views as *logical* idealism, as opposed to a subjective

---

[2]Maybe Cassirer's clearest expression of this relation is the following: 'If the real is represented as the results of the interpenetration of elementary series of dependencies, then in principle it has gained the form of a mathematically determinable structure' (SF, 257).

[3]Moreover, this is one of the aspects of Cassirer's thinking that went through profound changes after SF.

[4]Friedman [2000] contains a penetrating analysis of the problems this creates for Cassirer.

[5]'The constancy of the ideal forms has no longer a purely static, but also and especially a dynamic meaning; it is not so much constancy in *being*, as rather constancy in logic *use*' (SF, 323)

or mental idealism. He constantly warns against the confusion that arises if one forgets that the central problem for critical philosophy is 'knowing objectively' and not 'knowing objects'. Hence there is an important difference between claiming objective validity for certain procedures and asserting the existence of a class of objects.

This distinction, which is central to Cassirer's project, is highly relevant with respect to the problem of idealization. For this reason, Mormann's paper should be applauded for connecting Cassirer's philosophy with the concerns of contemporary philosophy of science about idealization, laws, and models. As everywhere Cassirer opposes forcefully to the metaphysical reification of logical procedures: 'the existence of the ideal, which can alone be critically affirmed and advocated, means nothing more than the objective logical necessity of idealization' (SF, 129). The ideal elements 'only go beyond the given, in order to grasp the more sharply the systematic structural relations of the given' (SF, 128).

Nowhere does Cassirer show interest in ascertaining the origins of the contents of our knowledge. His interest solely lies in determining the structure of knowledge and in isolating the logical elements which are posited by pure reason. Cassirer's idealism is not the metaphysical idealism of the nineteenth century German post-Kantian thinkers, but the epistemological and anti-metaphysical idealism of the neo-Kantians associated with the Marburg school.[6] In this respect his most original contribution may be the relativization of all oppositions between object and subject, universal and particular, content and form, arguing for a view in which the correlation between these opposing poles is the primary moment—but we will no go into these issues.

3 Ideal Elements and Empirical Reality

As mentioned in our introduction, we believe that Mormann could have strengthened his plea for a renewed attention to Cassirer by mentioning some further aspects of Cassirer's views which are of relevance for contemporary philosophy of science. These aspects have to do with the *application* of the concepts constructed by the method of ideal elements to empirical reality. His views on this application come surprisingly close to some contemporary ideas.

3.1

Whereas the first part of SF is concerned with the problem of concept formation, the second part tackles 'the problem of reality'. The first part had to remain incomplete from the epistemological point of view since it only dealt with the a priori aspects of knowledge, and the second part thus provides a necessary completion.

---

[6]For a useful sketch of the landscape of neo-Kantian philosophy in the beginning of the twentieth century, see the third chapter of [Friedman, 2000].

Cassirer continually stresses that physical laws will always be true of their concepts, since they are constructed in that way. However, this does not guarantee their applicability to concrete situations. An empirical moment has to supplement the purely conceptual determination: 'The preliminary deductive work furnishes a survey of the possible kinds of exact correlation; while experience determines which of the possible types of connection is applicable to the case in hand' (SF, 150).[7] This is not all for Cassirer, as he adds that empirical reality is in principle inexhaustible. As an example he adduces the problem of the fall of sensuously real bodies, for the description of which progressively complex determinations that were originally excluded have to be added (e.g. the variation of the acceleration according to the distance from the centre of the earth, retardation by the resistance of the air, etc.) (SF, 254). Every determination necessarily remains preliminary, but this does not detract from the inner necessity of each determination.

> A contradiction here [between theoretical construction and actual observations] would not be resolved by giving up the principles of earlier investigations, but by adding new factors to these investigations, which correct the first result yet enable us to retain it in a new meaning. In so far as we abstract from subjective errors of observation, the *truth* of the individual determinations remains unaffected. What is always being questioned anew is only the *sufficiency* of these determinations for explaining the complicated factual relations of reality (SF, 254–5).

> The question is not whether to a strictly defined content a the predicate b belongs or not, but whether a [empirically] given content satisfies all the conditions of the concept a, or is to be determined by a different concept a'. The problem is not whether a is truly b, but whether x, which is offered by mere perception, is truly a. (244–5)

3.2

Cassirer's account of the application of scientific theories to concrete problems thus shows a two-tiered structure. There is the construction of the ideal concepts, which receive their meaning from the (mathematical) structures in which they are embedded, and there is the identification of empirical situations with conceptual determinations. This latter identification will always be only approximately valid, and will possibly result in the addition of extra conceptual determinations.

Compare this with the account given by Ronald Giere in his *Explaining Science*, one of the most influential statements of the model-theoretic view on scientific theories. Giere explains that any adequate theory of scientific practice must treat two distinct phenomena:

---

[7]This should also counter all remaining suspicions about the nature of Cassirer's idealism.

One is the linking of the mathematical symbols with *general terms*, or concepts, such as 'position'. A second phenomenon is the linking of a mathematical symbol with some feature of a *specific object*, such as 'the position of the moon'. I will refer to the former as the problem of *interpretation* and the latter as the problem of *identification* [Giere, 1988, p. 75].

Even more strikingly, Giere claims that laws will (by definition) always be true of the corresponding models. Moreover, he also explains how the identification will always be only approximately valid and he shows how scientists deal with this by the addition of correction terms. Probably not accidentally, Mormann's reformulation of the method of ideal systems in terms of state spaces does also point towards a central aspect of the model-theoretic view. Let us suffice with remarking that Mormann is clearly right when claiming that Cassirer's work 'might be of some interest for the contemporary agenda of philosophy of science'. [8]

4 Moderate Conventionalism?

Mormann suggests to conceive Cassirer's idealist philosophy of science as a kind of moderate conventionalism. We will not enter into nominal disputes, and we readily acknowledge conventionalist tendencies in Cassirer's thought (who at different places cites Poincaré approvingly), but we want to point to some differences between Cassirer and Poincaré. Among other things, this will help to bring out the sophistication of Cassirer's views, since without doubt Poincaré is one of the most respected philosophers of mathematics. At the same time, this will serve as a warning against too hasty attempts at classifying Cassirer in terms familiar for analytic philosophy of science.

4.1

Henri Poincaré lived from 1854 till 1912. Cassirer was 20 years younger (1874–1945), but their work on the philosophy of mathematics dates from the same period: (roughly) the first decade of the 20th century. If we compare the two, the most striking similarity is that their view on geometry and its evolution in the 19th century is basically the same. According to Cassirer the developments in the 19th century make a Kantian intuition-oriented view on geometry untenable, as already explained. Poincaré arrives at the same conclusion and uses the same type of argument. Let us expand on this a bit.

One of the standard arguments against Kantian intuitionism is what Philip Kitcher calls the *practical impossibility problem* [Kitcher, 1984, p. 51]. The ex-

---

[8]There are important further connections that might be mentioned. Some of Cassirer's claims come surprisingly close to what is asserted by modern-day structural realists. An important difference nevertheless remains in the critical (transcendental) perspective of Cassirer's thought—not so much the nature of reality, but the nature of the necessary conditions of knowledge is at stake for him.

amples which Kant gives to illustrate how intuition is supposed to work, relate to geometry: he does not clarify how we have to construct mental images of numbers and sets, and derive properties from those images. But even within geometry there are problems. Kant claim e.g. that we can see 'by intuition' that line segments are infinitely divisible. However, there is no single mental construction which enables us to 'see' this; what is required is an infinite series of mental constructions, in which the given line segment is divided further and further.

The practical impossibility argument is a typical a priori argument. Poincaré uses an a posteriori argument against Kant, based on 'matters of fact' of mathematics as a discipline. He claims that the axioms of geometry cannot be synthetic a priori truths because, if they were, we could not conceive their negation, nor build theoretical systems on their negation: there would be no non-Euclidean geometries. Cassirer argues that since the role of intuition was continually altered in the development of modern geometry, its true foundations must lie somewhere else. So both Poincaré and Cassirer use a posteriori arguments,[9] although they disagree on the meaning of synthetic a priori judgements (as should be clear from our exposition of Cassirer's arguments in Section 2).

Before we have a look at the major differences between both thinkers, let us ask whether the a posteriori arguments are convincing. The least we can say is that Poincaré is a bit too quick: the existence of non-Euclidean geometry only entails that if there is a 'natural geometry' (i.e. a geometry which we necessarily use for the spatial ordering of our unstructured sensory inputs) we cannot know it: different people simultaneously have different intuitions about the content of this natural geometry, and we cannot know whose intuitions are the correct ones. Kitcher calls this the *exactness problem*. While Poincaré points at this exactness problem from a static, synchronic point of view (by insisting on coexisting axiomatizations of geometry), Cassirer points at the same problem in a dynamic, diachronic way (by exposing the evolution in the concepts of geometry). So their arguments are complementary.

4.2

Poincaré, in contradistinction to Cassirer, adopts Kantian intuitionism with respect to arithmetic. For instance, he claims that the axiom of induction (which he also calls the 'rule of reasoning by recurrence') is an a priori synthetic proposition. The following quotes from the first chapter of *Science and Hypothesis* (taken from the reprint in [Benacerraf and Putnam, 1983]) are clear enough:

> We cannot therefore escape the conclusion the rule of reasoning by recurrence is irreducible to the principle of contradiction. Nor can the rule come to us from experiment.

---

[9]Although it needs to be mentioned that at some places in SF Cassirer also hints at a practical impossibility argument.

...

This rule, inaccessible to analytical proof and to experiment, is the exact type of the *a priori* synthetic intuition. On the other hand, we cannot see in it a convention as in the case of the postulates of geometry.

...

Induction applied to the physical sciences is always uncertain, because it is based on the belief in a general order of the universe, an order which is external to us. Mathematical induction—i.e., proof by recurrence—is, on the contrary, necessarily imposed on us, because it is only the affirmation of the property of the mind itself. [Benacerraf and Putnam, 1983, pp. 400–401]

Given his synchronic approach and the fact that the axiomatization of arithmetic was in an early stage at that time (Peano's *Arithmetices Principia Nova Methodo Exposita* was published in 1889), this position is understandable: at the time of Poincaré there was only Peano's axiomatization. The evolution of arithmetic has shown that Poincaré was wrong in his judgement: there can be arithmetics that do without the axiom of induction (e.g. by replacing Peano's second order induction axiom with a first order axiom scheme) and other non-Peano arithmetics (e.g. finitistic arithmetics, which violate other axioms of Peano's system).

Cassirer does not attribute a different status to arithmetic and geometry. Again this can be understood from the way he argues. Evolutions of the kind by which Cassirer shows that there is an exactness problem, occur both in geometry and in arithmetic. Moreover, this is clearly connected with his overall philosophical program, in which he tried to do entirely without the pure forms of sensible intuition, following his teacher Hermann Cohen.

4.3

We believe that Cassirer's attention to the dynamic character of thought is one of the aspects of his thought that clearly distinguishes him from Poincaré's conventionalism. This is in particular so because he also stresses that the evolution of intellectual constructions is inherently progressive and is supposed to converge on the unattainable limit of the perfectly adapted conceptual system. This may sound naïve to post-Kuhnian ears, but it needs to be remembered that in Cassirer's critical philosophy there are good reasons for demanding such a convergence, as he states himself: 'the assumption of this limit is not arbitrary, but inevitable, since only by it is the continuity of experience established' (SF, 322). Moreover—but this is only a suggestion waiting future elaboration—the reasons behind this demand, which are central to critical philosophy, might be used to throw some light on the vexing problems surrounding incommensurability. [10]

A last respect, and maybe the most important, in which Cassirer's critical idealism differs from Poincaré's conventionalism, is the constitutive role played by the 'free constructions'. Whereas Poincaré held on to an epistemologically supreme role for 'brute facts', and reserved the conventionality for the language in which these facts are expressed, Cassirer denies such an absolute distinction. (Still, it needs to be stressed that such a distinction has relative validity for Cassirer—this is important, because otherwise the elements of conventionalism that are present in his thinking would threaten to collapse his critical idealism into a complete relativism.) Moreover, in his view something can only become factual if it has undergone synthetisising conceptual determinations. We would only distort Cassirer's philosophy beyond recognition if we severed it from its transcendental origins.

4.4

In the end the fundamental difference in outlook between Poincaré and Cassirer might be traced back to the former's position as a practising scientist, for whom the real interest in understanding the foundations of his discipline ultimately lies in the possible future directions of work. Cassirer's project on the other hand is an investigation into the general structure of knowledge, a task for which he must question certain natural assumptions made by all scientists (e.g. on the nature of 'facts').

> [C]ritical thought is not directed forwards on the gaining of new objective experiences, but backwards on the origin and foundations of knowledge. The two tendencies of thought here referred to can never be directly united; the conditions of scientific *production* are different from those of critical *reflection*. ... The peculiar character of knowledge rests on the tension and opposition remaining between these standpoints (SF, 210).

Acknowledgements

Maarten Van Dyck is a Research Assistant of the Fund for Scientific Research, Flanders.

BIBLIOGRAPHY

[Benacerraf and Putnam, 1983] P. Benacerraf and H. Putnam (eds.). *Philosophy of Mathematics. Selected Readings*. Cambridge: Cambridge University Press, 1983.
[Cassirer, 1907] E. Cassirer. Kant und die moderne Mathematik. *Kant-Studien*, XII, 1–49, 1907.

---

[10]Particularly interesting in this respect would be a confrontation of Cassirer's philosophy with Paul Hoyningen-Huene's neo-Kantian reconstruction of Kuhn's philosophy of science in [Hoyningen-Huene, 1993]. However, this task would also be complicated by the developments in Cassirer's thought after SF; complications which nevertheless might also prove instructive.

[Cassirer, 1953] E. Cassirer. *Substance and Function & Einstein's Theory of Relativity*. New York: Dover, 1953.

[Friedman, 2000] M. Friedman. *A Parting of the Ways. Carnap, Cassirer, and Heidegger*. Peru (Ill.): Open Court, 2000.

[Giere, 1988] R. N. Giere. *Explaining Science. A Cognitive Approach*. Chicago and London: University of Chicago Press, 1988.

[Hoyningen-Huene, 1993] P. Hoyningen-Huene. *Reconstructing Scientific Revolutions*. Chicago and London: The University of Chicago Press. 1993.

[Kitcher, 1984] P. Kitcher. *The Nature of Mathematical Knowledge*. New York and Oxford: Oxford University Press, 1984.

[Krois, 1987] J. M. Krois. *Cassirer. Symbolic Forms and History*. New Haven and London: Yale University Press, 1987.

[Passmore, 1966] J. Passmore. *A Hundred Years of Philosophy*. Duckworth, London, 1966.

Maarten Van Dyck and Erik Weber

Centre for Logic and Philosophy of Science
Ghent University, Blandijnberg 2,
B-9000 Gent, Belgium

Email: {maarten.vandyck, erik.weber}@UGent.be

Recursive Causality in Bayesian Networks and Self-Fibring Networks

Jon Williamson and Dov Gabbay

Causal relations can themselves take part in causal relations. The fact that smoking causes cancer (SC), for instance, causes government to restrict tobacco advertising (A), which helps prevent smoking (S), which in turn helps prevent cancer (C). This causal chain is depicted in Figure 1, and further examples will be given in Section 1.

Figure 1. SC: smoking causes cancer; A: tobacco advertising; S: smoking; C: cancer

So causal models need to be able to treat causal relationships as causes and effects. This observation motivates an extension the Bayesian network causal calculus (Section 2) to allow nodes that themselves take Bayesian networks as values. Such networks will be called *recursive Bayesian networks* (Section 3).

Because recursive Bayesian networks make causal and probabilistic claims at different levels of their recursive structure, there is a danger that the network might contradict itself. Hence we need to ensure that the network is consistent, as explained in Section 4. Having done this, in Section 5 we propose a new Markov condition: under this condition a recursive Bayesian network determines a joint probability distribution over its domain.

In Section 6 we compare our approach to other generalisations of Bayesian networks, and in Section 7 we show by analogy with recursive Bayesian networks how recursive causality can be modelled in structural equation models. A similar analogy motivates the application of recursive Bayesian networks to a non-causal domain, namely the modelling of arguments (Section 8).

A recursive Bayesian network is an instance of a very general structure called a self-fibring information network, whose properties are explored in Section 9 and Section 10.

Laws and Models in in Science, 173–221.
© *2004, the author.*

1   Causal Relations as Causes

It is almost universally accepted that causality is an asymmetric binary relation. [1]
But the question of what the causal relation relates is much more controversial:
the relata of causality have variously taken to be single-case events, properties,
propositions, facts, sentences and more. In this paper we shall only add to the controversy, by dealing with cases in which causal relations themselves are included
as relata of causality. Our aim is to shed light more on the processes of causal
reasoning, especially formalisations of causal reasoning, than on the metaphysics
of causality.

More generally we shall consider sets of causal relations, represented by directed causal graphs such as that of Figure 1, as relata of causality. (A single
causal relationship is then represented by a causal graph consisting of two nodes
referring to the relata and an arrow from cause to effect.) If, as in Figure 1, a causal
graph G contains a causal relation or causal graph as a value of a node, we shall
call G a *recursive causal graph* and say that it represents *recursive causality*.

Perhaps the best way to get a feel for the importance and pervasiveness of recursive causality is through a series of examples.

Policy decisions are often influenced by causal relations. As we have already
seen, smoking causing cancer itself causes restrictions on advertising. Similarly,
monetary policy makers reduce interest rates (R) because interest rate reductions
boost the economy (E) by causing borrowing increases (B) which in turn allow
investment (I). Here we have a causal chain as in Figure 2 forming the value of
node RE in Figure 3.

Figure 2. R: interest rate reduction; B: borrowing; I: investment; E: economic
boost

Figure 3. RE: interest rate reduction causing economic boost; R: interest rate
reduction

Policy need not be made for us: we often decide how we behave on the basis
of perceived causal relationships. It is plausible that drinking red wine causes an

---

[1]Not quite universally: [Mellor, 1995] disagrees for example.

increase in anti-oxidants which in turn reduces cholesterol deposits, and this apparent causal relationship causes some people to increase their red wine consumption.
This example highlights two important points. Firstly, it is a belief in the causal
relationship which directly causes the policy change, not the causal relationship itself. The belief in the causal relationship may itself be caused by the relationship,
but it may not be—it may be a false belief or it may be true by accident. Likewise,
if a causal relationship exists but no one believes that it exists, there will be no
policy change. Secondly, the policy decision need not be rational on the basis of
the actual causal relationship that causes the decision: drinking red wine may do
more harm than good.

A contract can be thought of as a causal relationship, and the existence of a
contract can be an important factor in making a decision. A contract in which
production of commodity C is purchased at price P may be thought of as a causal
relationship $C \longrightarrow P$, and the existence of this causal relationship can in turn
cause the producer to invest in further means of production, or even other com-
modities. For example, a Fair Trade chocolate company has a long-term contract
with a co-operative of Ghanaian cocoa producers to purchase (P) cocoa (C) at a
price advantageous to the producer as in Figure 4. The existence of this contract
(CP) allows the co-operative to invest in community projects such as schools (S),
as in Figure 5.

Figure 4. C: cocoa production; P: purchase

Figure 5. CP: cocoa production causing payment; S school investment

An insurance contract is an important instance of this example of recursive
causality. Insuring a building against fire may be thought of as a causal relationship of the form 'insurance contract causes [fire F causes remuneration R]' or
$[C \longrightarrow P] \longrightarrow [F \longrightarrow R]$ for short, where as before C is the commodity (i.e. the
contract) and P is payment of the premium. The existence of such an insurance
policy can cause the policy holder to commit arson (A) and set fire to her building
and thereby get remunerated: $[[C \longrightarrow P] \longrightarrow [F \longrightarrow R]] \longrightarrow A \longrightarrow F \longrightarrow R$.
Causality in this relationship is nested at three levels. Insurance companies will
clearly want to limit the probability of remuneration given that arson has occurred.

Thus we see that recursive causality is particularly pervasive in decision-making scenarios. However, recursive causality may occur in other situations too—situations in which it is the causal relationship itself, rather than someone's belief in the relationship, that does the causing. Pre-emption is an important case of recursive causality, where the pre-empting causal relationship prevents the pre-empted relationship: [poisoning causing death] prevents [heart failure causing death].[2] Context-specific causality may also be thought of recursively: a causal relationship that only occurs in a particular context (such as susceptibility to disease amongst immune-deficient people) can often be thought of in terms of the context causing the causal relationship.

Arguably prevention is often best interpreted in terms of recursive causality: when taking mineral supplements prevents goitre, what is really happening is that taking mineral supplements prevents [poor diet causing goitre]—this is because there are other causes of goitre such as various defects of the thyroid gland, taking mineral supplements does not inhibit these causal chains and thus does not prevent goitre simpliciter. (In many such cases, however, the recursive nature can be eliminated by identifying a particular component of the causal chain which is prevented. Poor diet (D) causes goitre (G) via iodine deficiency (I) and mineral supplements (S) prevent iodine deficiency and so this example might be adequately represented by Figure 6, which is not recursive. Of course the recursive aspect can not be eliminated if no suitable intermediate variable I is known to the modeller.)

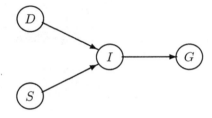

Figure 6. D: poor diet; S: mineral supplements; I: iodine deficiency; G: goitre

Recursive causality is clearly a widespread phenomenon. The question now arises as to how recursive causality ought to influence our reasoning mechanisms. After a brief introduction to Bayesian networks in Section 2 we shall extend the Bayesian network formalism to cope with recursive causality (Section 3) and then discuss some related extensions of Bayesian networks (Section 6). Later we shall see that this approach to causal reasoning generalises in an interesting way.

---

[2]We suggest that this is a simpler and more natural way of representing pre-emption than the proposal of Section 10.1.3, 10.3.3, 10.3.5 of [Pearl, 2000].

2   Bayesian Networks

A Bayesian network is defined over a finite domain $V = \{V_1, \ldots, V_n\}$ of variables. In principle there are no size restrictions on the set of possible values that each variable may take, but often in practice each variable will have only a finite number of possible values. For simplicity we shall restrict our attention to two-valued variables, and denote the assignment of V_i to its values by v_i and $\neg v_i$ respectively, for $i = 1, \ldots, n$. An assignment u to a subset $U \subseteq V$ of variables is a conjunction of assignments to each of the variables in U. For example $v_1 \wedge \neg v_2 \wedge \neg v_5$ is an assignment to $\{V_1, V_2, V_5\}$.

A (causally interpreted) *Bayesian network* b on V consists of two components:

- A directed acyclic graph G with nodes from V, representing the causal relations amongst the variables.

- A probability specification S. For each $V_i \in V$, S specifies the probability distribution of V_i conditional on its parents (direct causes in G), i.e. S consists of statements of the form '$p(v_i|par_i) = x_{i,par_i}$' for each $i = 1, \ldots, n$ and assignment par_i of values to the parents of V_i, and where each $x_{i,par_i} \in [0,1]$. If the value of a variable V_i is known then V_i is said to be *instantiated* to that value and the corresponding probability specifiers $p(v_i|par_i)$ are 1 or 0 according to whether v_i or $\neg v_i$ is the instantiated value.

The graph and probability specification of a Bayesian network are linked by a fundamental assumption known as the *causal Markov condition*. This says that conditional on its parents, any node is probabilistically independent of all other nodes apart from its descendants, written $V_i \perp\!\!\!\perp ND_i \mid Par_i$ where ND_i and Par_i are respectively the sets of non-descendants and parents of V_i.

A Bayesian network suffices to determine a joint probability distribution over its nodes, since for each assignment v on V,

$$(1) \quad p(v) = \prod_{i=1}^{n} p(v_i|par_i)$$

where v_i is the assignment v gives to V_i, and par_i is the assignment v gives to the parents Par_i of V_i.

Bayesian networks are used because they offer the opportunity of an efficient representation of a joint probability distribution over V. While 2^n different probabilities $p(v)$ specify the joint distribution, these values may (depending on the structure of the causal graph G) be determined via eqn 1 from relatively few values in the probability specification S. Furthermore, a number of algorithms have been developed for determining marginal probabilities from a Bayesian network, often very quickly (but this again depends on the structure of G).[3] Causal graphs are often sparse, and thus lead to efficient Bayesian network representations. Moreover the causal interpretation of the graph ensures that the causal Markov condition is a good default assumption, even if the conditional independence relationships it posits do not always hold in practice.[4]

3   Extension to Recursive Causality

As noted in Section 1, causal relationships often act as causes or effects themselves. In a Bayesian network, however, the nodes tend to be thought of as simple variables, not complex causal relationships. Thus we need to generalise the concept of Bayesian network so that nodes in its causal graph G can signify complex causal relationships. On the other hand, we would like to retain the essential features of ordinary networks, namely the ability to represent joint distributions efficiently, and the ability to perform probabilistic inference efficiently.

The essential step is this. We shall allow variables to take Bayesian networks as values. If a variable takes Bayesian networks as values we will call it a *network variable* to distinguish it from a *simple variable* whose values do not contain such structure. Thus S, which signifies 'payment of subsidy to farmer' and takes value true (s) or false ($\neg s$) is a simple variable. But an example of a network variable is A, which stands for 'agricultural policy' and takes value a signifying the Bayesian network containing the graph of Figure 7 and the specification $\{p_a(f) = 0.1, p_a(s|f) = 0.9, p_a(s|\neg f) = 0.2\}$, where F is a simple variable signifying 'farming', or value $\neg a$ signifying Bayesian net with graph of Figure 8 and specification $\{p_{\neg a}(f) = 0.1, p_{\neg a}(s) = 0.2\}$. Here a is a policy in which farming causes subsidy and $\neg a$ is a policy in which there is no such causal relationship. For simplicity we shall consider network variables with at most two values, but the theory that follows applies to network variables which take any finite number of values.

Figure 7. Graph of a: farming causes subsidy

---

[3]See [Neapolitan, 1990] for a detailed discussion of the properties of Bayesian networks and key inference algorithms.
    [4]See [Williamson, 2001] on this point. While Bayesian networks were originally developed with a causal interpretation in mind [Pearl, 1988], a joint probability distribution can also be represented by a Bayesian network whose graph does not admit a viable causal interpretation. If a Bayesian network is not causally interpreted then causal justifications of the Markov condition do not apply, and an independent justification is required. Thus in Section 5 we define a network called a *flattening* which contains arrows that do not correspond to causal relations, and we also provide a justification for the Markov condition.

Figure 8. Graph of ¬a: no causal relationship between farming and subsidy

A *recursive Bayesian network* is then a Bayesian network containing at least one network variable. For example the network with graph Figure 9 and specification $\{p(l) = 0.7, p(a|l) = 0.95, p(a|\neg l) = 0.4\}$, representing the causal relationship between lobbying and agricultural policy, is a recursive Bayesian network, where the simple variable L stands for 'lobbying' and takes value true or false, and A is the network variable signifying 'agricultural policy' discussed above.

Figure 9. Lobbying causes agricultural policy

We shall allow network variables to take recursive Bayesian networks (as well as the standard Bayesian networks of Section 2) as values. In this way a recursive Bayesian network represents a hierarchical structure.

If a variable C is a network variable then the variables that occur as nodes in the Bayesian networks that are the values of C are called the *direct inferiors* of C, and each such variable has C as a *direct superior*. *Inferior* and *superior* are the transitive closures of these relations: thus E is inferior to C iff it is directly inferior to C or directly inferior to a variable D that is inferior to C. The variables that occur in the same local network as C are called its *peers*.

A recursive Bayesian network $b = (G, S)$ conveys information on a number of levels. The variables that are nodes in G are *level 1*; any variables directly inferior to level 1 variables are *level 2*, and so on. The network b itself can be associated with a network variable B that is instantiated to value b, and we can speak of B as the *level 0 variable*. (We have not specified the other possible values of B: for concreteness we can suppose that B is a single-valued network variable which only takes value b.) The *depth* of the network is the maximum level attained by a variable. A Bayesian network is *non-recursive* if its depth is 1; it is *well-founded* if its depth is finite. We shall restrict our discussion to *finite* networks: well-founded networks whose levels are each of finite size.

For $i \geq 0$ let V_i be the set of level i variables, and let $\mathcal{G}_i$ and $\mathcal{S}_i$ be the set of graphs and specifications respectively that occur in networks that are values of level i variables. Thus $V_0 = \{B\}, \mathcal{G}_0 = \{G\}$ and $\mathcal{S}_0 = \{S\}$. The domain of b is the set $V = \bigcup_i V_i$ of variables at all levels.

Note that V contains the level 0 variable B itself and thus contains all the structure of b. In our example $V = \{B, L, A, F, S\}$ where the level 0 network variable B takes value b whose graph is Figure 9 and whose probability specification is $\{p(l) = 0.7, p(a|l) = 0.95, p(a|\neg l) = 0.4\}$ and the only other network variable is A whose value a has graph of Figure 7 and specification $\{p_a(f) = 0.1, p_a(s|f) = 0.9, p_a(s|\neg f) = 0.2\}$ and whose value $\neg a$ has graph of Figure 8 and specification $\{p_{\neg a}(f) = 0.1, p_{\neg a}(s) = 0.2\}$; then V itself determines all the structure of the recursive Bayesian network in question. Consequently we can talk of 'recursive Bayesian network b on domain V' and 'recursive Bayesian network of V' interchangeably.

A network variable V_i can be thought of as a simple variable V_i' if one drops the Bayesian network interpretation of each of its values: V_i' is the *simplification* of V_i. A recursive network b can then be interpreted as a non-recursive network b' on domain $\mathcal{V}_1' = \{V_i' : V_i \in \mathcal{V}_1\}$: then b' is called the *simplification* of b.

A variable may well occur more than once in a recursive Bayesian network, in which case it might have more than one level.[5] Note that in a well-founded network no variable can be its own superior or inferior. A recursive Bayesian network makes causal and probabilistic claims at all its various levels, and if variables occur more than once in the network, these claims might contradict each other. We shall examine this possibility now.

4   Consistency

Network variables that occur in the domain of a recursive Bayesian network $b = (G, S)$ can be interpreted as making causal and probabilistic claims about the world. Any network variable that is instantiated to a particular value asserts the validity of the network to which it is instantiated. In particular the level 0 network variable B asserts its instantiated value b, i.e. it asserts the causal relations in G, the probabilistic independence relationships one can derive from G via the causal Markov condition, and the probabilistic claims made by the probability specification S. A network variable that is not instantiated asserts the weaker claim that precisely one of its possible values is correct. A recursive Bayesian network is consistent if these claims do not contradict each other.

In order to give a more precise formulation of the consistency requirement we need first to define consistency of non-recursive Bayesian networks. There are three desiderata: consistency with respect to causal claims (*causal consistency*), consistency with respect to implied probabilistic independencies (*Markov consis-*

---

[5]While one might think that there will be no repetition of variables if all variables correspond to single-case events, this is not so. Event A causing event B causes an agent to change her belief about the relationship between A and B, this belief being represented by network variable C whose value $\neg c$ has B causing A and whose value c has A causing B. Here A and B occur more than once in the network but need not be repeatably instantiatable variables—they may be single-case events.

tency) and consistency with respect to probabilistic specifiers (*probabilistic consistency*).

First causal consistency. A *chain* $A \rightsquigarrow B$ from node A to node B in a directed acyclic graph is a sequence of nodes in the graph, beginning with A and ending with B, such that there is an arrow from each node to its successor. A *subchain* of a chain c from A to B is a chain from A to B involving nodes in c in the same order, though not necessarily all the nodes in c. Thus Figure 10 contains both the chain (A, C, B) and its subchain (A, B). The *interior* of a chain $A \rightsquigarrow B$ is defined as the subchain involving all nodes between A and B in the chain, not including A and B themselves. A *path* between A and B is a sequence of nodes in the graph, beginning with A and ending with B, such that there is an arrow between each node and its successor (the direction of the arrow is unimportant).

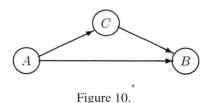

Figure 10.

The restriction $G_{\downarrow W}$ of causal graph G defined on variables V to the set of variables $W \subseteq V$ is defined as follows: for variables $A, B \in W$, there is an arrow $A \longrightarrow B$ in $G_{\downarrow W}$ if and only if $A \longrightarrow B$ is in G or, $A \rightsquigarrow B$ is in G and the variables in the interior of this chain are in $V \backslash W$. Thus G and $G_{\downarrow W}$ agree as to the causal relationships amongst variables in W. It is not hard to see that for $X \subseteq W \subseteq V, G_{\downarrow W \downarrow X} = G_{\downarrow X}$.

Two causal graphs G on V and H on W are *causally consistent* if there is a third (directed and acyclic) causal graph F on $U = V \cup W$ such that $F_{\downarrow V} = G$ and $F_{\downarrow W} = H$. Thus G and H are causally consistent if there is a model F of the causal relationships in both G and H. Such an F is called a *causal supergraph* of G and H.

Figure 11 and Figure 12 are causally consistent for instance, because the latter graph is the restriction of the former to $\{A, B, C\}$. However, Figure 10 is not causally consistent with Figure 11: they do not agree as to the causal chains between A, B and C. Similarly Figure 10 and Figure 12 are causally inconsistent.

Figure 11.

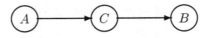

Figure 12.

Note that if G and H are causally consistent and nodes A and B occur in both G and H then there is a chain $A \rightsquigarrow B$ in G iff there is a chain $A \rightsquigarrow B$ in H. We will define two non-recursive Bayesian networks to be *causally consistent* if their causal graphs are causally consistent.

Another important consistency requirement is Markov consistency. Two causal graphs G and H are *Markov consistent* if they posit (via the causal Markov condition) the same set of conditional independence relationships on the nodes they share. Figure 11 and Figure 12 are Markov consistent because on their shared nodes A, C, B they each imply just that A and B are probabilistically independent conditional on C. Figure 10 is not Markov consistent with either of these graphs because it does not imply this independency. Two non-recursive Bayesian networks are *Markov consistent* if their causal graphs are Markov consistent.

Note that Markov consistency does not imply causal consistency: for instance two different complete graphs on the same set of nodes (graphs, such as Figure 10, in which each pair of nodes is connected by some arrow) are Markov consistent, since neither graph implies any independence relationships, but causally inconsistent because where they differ, they differ as to the causal claims they make. Neither does causal consistency of a pair of causal graphs imply Markov consistency: Figure 13 and Figure 14 are causally consistent but Figure 14 implies that A and B are probabilistically independent, while Figure 13 does not.

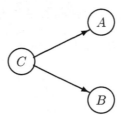

Figure 13.

In fact we have the following. Let $Com_G(X)$ be the set of closest common causes of X according to G, that is, the set of causes C of X that are causes of at least two nodes A and B in X for which some pair of chains from C to A and C to B only have node C in common. Then,

Figure 14.

PROPOSITION 1. *Suppose G and H are causal graphs on V and W respectively. G and H are Markov consistent if they are causally consistent and their shared nodes are closed under closest common causes ('cccc' for short), $Com_G(V \cap W) \cup Com_H(V \cap W) \subseteq V \cap W$.*

Proof. Suppose $X \perp\!\!\!\perp_G Y \mid Z$ for some $X, Y, Z \subseteq V \cap W$. Then for each $A \in X$ and $B \in Y$, Z *D-separates* A from B in G: every path between A and B is *blocked* by Z, i.e. every path contains (i) a structure $\longrightarrow C \longrightarrow$ with C in Z, or (ii) a structure $\longleftarrow C \longrightarrow$ with C in Z, or (iii) a structure $\longrightarrow C \longleftarrow$ where Z contains neither C nor any of its descendants.[6] G and H are causally consistent so there is a causal supergraph F on $V \cup W$ ($G = F_{\mid V}$ and $H = F_{\mid W}$). Now consider a path between A and B in F. Such a path either (a) is a chain ($A \rightsquigarrow B$ or $B \rightsquigarrow A$), (b) contains some C where $C \rightsquigarrow A$ and $C \rightsquigarrow B$, or (c) contains a $\longrightarrow C \longleftarrow$ structure. In case (a) there must be in G a subchain of this chain which is blocked by Z so the original chain in F must also be blocked by Z. Similarly in case (b), since G and H are cccc there must be a blocked subpath in G which has $C \rightsquigarrow A$ and $C \rightsquigarrow B$. In case (c), either there is a corresponding subpath in G which is blocked, or C and its descendants are not in Z so the path in F is blocked in any case. Thus $X \perp\!\!\!\perp_F Y \mid Z$. Next take the restriction $F_{\mid W} = H$. Paths between A and B in H must be blocked by Z since they are subpaths of paths in F that are blocked by Z and all variables in Z occur in H. Thus $X \perp\!\!\!\perp_H Y \mid Z$, as required. ∎

Note that while (under the assumption of causal consistency) closure under closest common causes is a sufficient condition for Markov consistency, it is not a necessary condition: Figure 13 and Figure 15 are Markov consistent because neither imply any independencies just amongst their shared nodes A and B, but the set of shared nodes is not closed under closest common causes.

---

[6]D-separation is a necessary and sufficient condition for deciding the conditional independencies implied by a causal graph under the causal Markov condition. See[Pearl, 1988, Section 3.3.1].

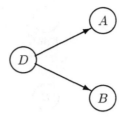

Figure 15.

Markov consistency is quite a strong condition. It is not sufficient merely to require that the pair of causal graphs imply sets of conditional independence relations that are consistent with each other—in fact any two graphs satisfy this property. The motivation behind Markov consistency is based on the fact that a cause and its effect are usually probabilistically dependent conditional on the effect's other causes (this property is known as the *causal dependence condition*), in which case probabilistic independencies that are not implied by the causal Markov condition are unlikely to occur. For example, while the fact that C causes A and B (Figure 13) is consistent with A and B being unconditionally independent (Figure 14), it makes their independence extremely unlikely: if A and B have a common cause then the occurrence of assignment a of A may be attributable to the common cause which then renders b more likely (less likely, if the common cause is a preventative), in which case A and B are unconditionally dependent. Thus Figure 13 and Figure 14 are not compatible, and we need the stronger condition that independence constraints implied by each graph should agree on the set of nodes that occur in both graphs.

Finally we turn to probabilistic consistency. Two causally consistent non-recursive Bayesian networks (G, S) and (H, T), defined over V and W respectively, are *probabilistically consistent* if there is some non-recursive Bayesian network (F, R), defined over $V \cup W$ and where F is a causal supergraph of G and H, whose induced probability function satisfies all the equalities in $S \cup T$. Such a network is called a *causal supernet* of (G, S) and (H, T).

PROPOSITION 2. *Suppose two non-recursive Bayesian networks (G, S) and (H, T) are causally consistent, probabilistically consistent and closed under closest common causes (cccc). Then there is a causal supernet (F, R) of (G, S) and (H, T) that is cccc with (G, S) and (H, T).*

Proof. Because (G, S) and (H, T) are causally and probabilistically consistent, there is a supernet (E, Q), of (G, S) and (H, T). If E is cccc with G and H then we set $(F, R) = (E, Q)$ and we are done. Otherwise, if E is not cccc with G say, then there is some Y-structure of the form of Figure 16 in E, where Figure 17 is the corresponding structure in G. (In these diagrams take the arrows to signify the existence of causal chains rather than direct causal relations.) Note that B must be in G or H, since the domain of a causal supergraph of G and H is the union of the domains of G and H; B cannot be in G since otherwise by causal consistency the chain from A to C in G would go via B; hence B is in H. Note also that not both of C and D can be in H, for otherwise G and H are not cccc. Suppose then that D is not in H. Then the chain from B to D is not in G or H. Construct F by taking E, removing the chain from B to D and including a chain from A to D, as in Figure 18. (Do this for all such Y-structures not replicated in G.) F remains a causal supergraph of G and H, since the chain from B to H was redundant. Moreover F is now cccc with G. Next construct the associated probability specification R by determining specifiers from (E, Q). Thus if the causal chain from A to D is direct we can set $p(d|a) = \sum_b p_{(E,Q)}(d|b)p_{(E,Q)}(b|a)$ in R. It is not hard to see that $p_{(F,R)}$ agres with $p_{(E,Q)}$ on the specifiers in S and T so the new network is also a causal supernet of (G, S). If E is not cccc with H then repeat this algorithm, to yield a causal supernet of (G, S) and (H, T) that is cccc with (G, S) and (H, T).

■

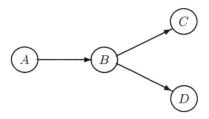

Figure 16. B is the closest common cause of C and D

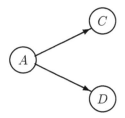

Figure 17. A is the closest common cause of C and D.

Note that the requirement that G and H are cccc in the above result is essential. If G is Figure 16 and H is Figure 17 then there is no causal supergraph of G and

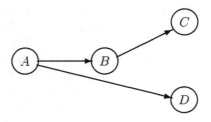

Figure 18. A is the closest common cause of C and D.

H that is cccc with G and H.

PROPOSITION 3. *Suppose two non-recursive Bayesian networks are causally consistent, probabilistically consistent and cccc. Then they determine the same probability function over the variables they share.*

Proof. Suppose (G, S) and (H, T) are causally and probabilistically consistent and cccc. Then by Proposition 2 there is a causal supernet (F, R) that is cccc with both nets. By Proposition 1 F is Markov consistent with G and H.

Next note that (G, S) and (F, R) determine the same probability function over variables V of (G, S):

$$p_{(G,S)}(v) = \prod_{v_i \in V} p_{(G,S)}(v_i | par_i^G)$$

where par_i^G is the state of the parents of V_i according to G that is consistent with assignment v to V,

$$= \prod_{v_i \in V} p_{(F,R)}(v_i | par_i^G)$$

since (F, R) is a causal supernet of (G, S),

$$= \prod_{v_i \in V} p_{(F,R)}(v_i | v_1, \ldots, v_{i-1}) = p_{(F,R)}(v)$$

where it is supposed that the variables $V_1, \ldots, V_n$ in V are ordered G-ancestrally, i.e. no descendants of V_i in G occur before V_i in the order. This last step follows because $V_i \perp\!\!\!\perp_G V_1, \ldots, V_{i-1} \mid Par_i^G$ implies $V_i \perp\!\!\!\perp_F V_1, \ldots, V_{i-1} \mid Par_i^G$ by Markov consistency.

Similarly (H, T) and (F, R) determine the same probability function over the variables of (H, T). Hence (G, S) and (H, T) determine the same probability function over variables they share.   ∎

Because Proposition 3 is a desirable property in itself we shall adopt closure under closest common causes as a consistency condition. We shall say that two non-recursive networks are *consistent* if they are causally and probabilistically consistent, and cccc. By Proposition 1 consistency implies Markov consistency.

Having elucidated concepts of consistency for non-recursive networks, we can now say what it means for a recursive network to be consistent.

An assignment v of values to variables in V, the domain of a recursive Bayesian network b, assigns values to all the simple variables and network variables that occur in b. Take for instance the recursive Bayesian network b of Figure 9: here $V = \{B, L, A, F, S\}$ and $b \wedge l \wedge \neg a \wedge f \wedge \neg s$ is an example of an assignment to V. (Note that the level 0 variable B only takes one value b and so must always be assigned this value.) Consider the assignment of values v gives to network variables in V. In our example, the network variables are B and A and these are assigned values b and $\neg a$ respectively. Each such value is itself a recursive Bayesian network, and when simplified induces a non-recursive Bayesian network. Let $\underline{b}_v$ denote the set of recursive Bayesian networks induced by v (i.e. the set of values v assigns to network variables of b) and let $\underline{b}'_v$ denote the set of non-recursive Bayesian networks formed by simplifying the networks in $\underline{b}_v$.

Assignment v is *consistent* if each pair of networks in $\underline{b}'_v$ is consistent (i.e. if each pair of values of network variables is consistent, when these values are interpreted non-recursively). A recursive Bayesian network is *consistent* if it has some consistent assignment v of values to V. A consistent assignment of values to the variables in a network can be thought of as a model or possible world, in which case consistency corresponds to satisfiability by a model.

In sum, if a recursive Bayesian network is not to be self-contradictory there must be some assignment under which all pairs of network variables satisfy three regularity conditions: causal consistency, probabilistic consistency and closure under closest common causes.

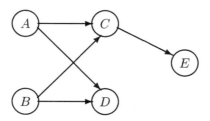

Figure 19. Graph G_1

Note that it is easy to turn a recursive network into one that is causally consistent, by ensuring that causal chains correspond for some assignment, and then cccc (and so Markov consistent), by ensuring that shared nodes of pairs of graphs also share closest common causes, for some assignment. In order to make G_2 in Figure 20 causally consistent with graph G_1 of Figure 19, for example, we need to introduce a chain that corresponds to the chain (D, F, E) in G_2, by adding an arrow from D to E in G_1. In order to make G_2 and G_1 cccc (and so Markov consistent) we need to add B to G_2 as a closest common cause of C and D. The modified graphs are depicted in Figure 21 and Figure 22.

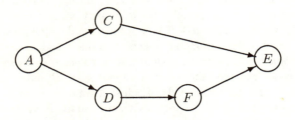

Figure 20. Graph G_2

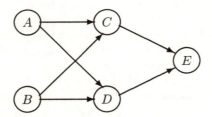

Figure 21. Graph H_1

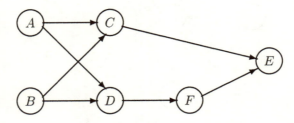

Figure 22. Graph H_2

Similarly in practice one would not expect each probability specification to be provided independently and then to have the problem of checking consistency—

ne would expect to use conditional distributions in one specification to determine distributions in others. For example, a probability specification on H_2 in Figure 22 would completely determine a probability specification on H_1 in Figure 21.

5   Joint Distributions

Any non-recursive Bayesian network is subject to the causal Markov condition (Section 2) which determines a joint probability distribution over the variables of the network from its graph and probability specification. We shall suppose that recursive Bayesian networks also satisfy the causal Markov condition. A recursive Bayesian network contains network variables whose values are interpreted as (recursive or non-recursive) Bayesian networks. Thus a recursive Bayesian network suffices to determine a hierarchy of joint probability distributions p_a on the (level 1) variables of a, for each a that occurs as the value of a network variable. (I.e. a recursive Bayesian network b determines a joint distribution on each network in $\underline{b}'_v$ for each consistent assignment v to the domain of b.) Standard Bayesian network algorithms can be used to perform inference in a recursive Bayesian network, and the range of causal-probabilistic questions that can be addressed is substantially increased. For example one can answer questions like 'what is the probability of a subsidy given farming?' (see Figure 7) and 'what is the probability of lobbying given agricultural policy $\neg a$?' (see Figure 9).

Certain questions remain unanswered however. We can not as yet determine the probability of one node conditional on another if the nodes only occur at different levels of the network. For example we can not answer the question 'what is the probability of subsidy given lobbying?' While we have a hierarchy of joint distributions, we have not yet specified a single joint distribution over the set of nodes in the union of the graph, i.e. over the recursive network as a whole.

In fact as we shall see, a recursive network does determine such an over-arching joint distribution if we make an extra independence assumption, called the *recursive Markov condition*: each variable is probabilistically independent of those other variables that are neither its inferiors nor its peers, conditional on its direct superiors.

A precise explication of the causal Markov condition and recursive Markov condition will be given shortly.

Given a recursive Bayesian network domain V and a consistent assignment v of values to V, we construct a non-recursive Bayesian network, the *flattening*, $v^\downarrow$, of v as follows. The domain of $v^\downarrow$ is V itself. The graph $G^\downarrow$ of $v^\downarrow$ has variables in V as nodes, each variable occurring only once in the graph. Add an arrow from V_i to V_j in $G^\downarrow$ if

- V_i is a parent of V_j in v (i.e. there is an arrow from V_i to V_j in the graph of some value of v) or

- V_i is a direct superior of V_j in v (i.e. V_j occurs in the graph of the value that v assigns to V_i).

We will describe the probability specification $S^\downarrow$ of $v^\downarrow$ in due course. First to some properties of the graph $G^\downarrow$.

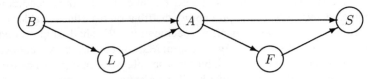

Figure 23. Example flattening

Note that $G^\downarrow$ may or may not be acyclic. If we take our farming example $V = \{B, L, A, F, S\}$ of Section 3, then the graph of the flattening $(b \wedge \neg l \wedge a \wedge f \wedge s)^\downarrow$ is depicted in Figure 23 and is acyclic. But the graph of the flattening of assignment $b \wedge c \wedge d \wedge e$ to $\{B, C, D, E\}$, where B is the level 0 network variable whose value b has graph $C \longrightarrow D$, C and E are simple variables and D is a network variable whose assigned value d has the graph $E \longrightarrow C$, is cyclic. The graph in a non-recursive Bayesian network must be acyclic in order to apply standard Bayesian network algorithms, and this requirement extends to recursive Bayesian networks: we will focus on consistent *acyclic* assignments to a recursive Bayesian network domain, those consistent assignments v that lead to an acyclic graph in the flattening $v^\downarrow$.[7]

By focussing on consistent acyclic assignments v, the following explications of the two independence conditions become plausible. Given a consistent acyclic assignment v, let PND_i^v be the set of variables that are peers but not descendants of V_i in v, NIP_i^v be the variables that are neither inferiors nor peers of V_i, and $DSup_i^v$ be the direct superiors of V_i. As before, Par_i^v are the parents of V_i and ND_i^v are the non-descendants of V_i. None of these sets are taken to include V_i itself.

Causal Markov Condition (CMC) For each $i = 1, \ldots, n$ and $DSup_i^v \subseteq X \subseteq NIP_i^v$, $V_i \perp\!\!\!\perp PND_i^v \mid Par_i^v, X$.

Recursive Markov Condition (RMC) For each $i = 1, \ldots, n$ and $Par_i^v \subseteq X \subseteq PND_i^v$, $V_i \perp\!\!\!\perp NIP_i^v \mid DSup_i^v, X$.

Then the graph of the flattening has the following property:

---

[7]Cyclic Bayesian networks have been studied to some extent, but are less tractable than the acyclic case: see [Spirtes, 1995] and [Neal, 2000].

PROPOSITION 4. *Suppose v is a consistent acyclic assignment to a recursive Bayesian network domain V. Then the probabilistic independencies implied by v via the causal Markov condition and the recursive Markov condition are just those implied by the graph $G^{\downarrow}$ of the flattening $v^{\downarrow}$ via the causal Markov condition.*

Proof. Order the variables in V ancestrally with respect to $G^{\downarrow}$, i.e. no descendants of V_i in $G^{\downarrow}$ occur before V_i in the ordering—this is always possible because $G^{\downarrow}$ is acyclic.

First we shall show that CMC and RMC for v imply CMC for $G^{\downarrow}$. By Corollary 3 of [Pearl, 1988] it suffices to show that $V_i \perp\!\!\!\perp V_1, \ldots, V_{i-1} \mid Par_i^{G^{\downarrow}}$ for any $V_i \in V$. By CMC, $V_i \perp\!\!\!\perp PND_i^v \mid Par_i^v, DSup_i^v$, and by RMC, $V_i \perp\!\!\!\perp NIP_i^v \mid DSup_i^v, PND_i^v$. Applying contraction,[8]

$$V_i \perp\!\!\!\perp PND_i^v \cup NIP_i^v \mid Par_i^v, DSup_i^v.$$

Now $\{V_1, \ldots, V_{i-1}\} \subseteq PND_i^v \cup NIP_i^v$ since the variables are ordered ancestrally and v is acyclic, and the parents of V_i in $G^{\downarrow}$ are just the parents and direct superiors of V_i in v, $Par_i^{G^{\downarrow}} = Par_i^v \cup DSup_i^v$, so $V_i \perp\!\!\!\perp V_1, \ldots, V_{i-1} \mid Par_i^{G^{\downarrow}}$ as required.

Next we shall see that CMC for $G^{\downarrow}$ implies CMC and RMC for v. In fact this follows straightforwardly by D-separation. $Par_i^v \cup X$ D-separates V_i and PND_i^v in $G^{\downarrow}$ for any $DSup_i^v \subseteq X \subseteq NIP_i^v$, since $Par_i^v \cup X$ includes the parents of V_i in $G^{\downarrow}$ and (by acyclicity of v) PND_i^v are non-descendants of V_i in $G^{\downarrow}$, so CMC holds. $DSup_i^v \cup X$ D-separates V_i and NIP_i^v in $G^{\downarrow}$ for any $Par_i^v \subseteq X \subseteq PND_i^v$, since $DSup_i^v \cup X$ includes the parents of V_i in $G^{\downarrow}$ and (by acyclicity of v) NIP_i^v are non-descendants of V_i in $G^{\downarrow}$, so RMC holds. ∎

Having defined the graph $G^{\downarrow}$ in the flattening $v^{\downarrow}$ of v, and examined its properties, we shall move on to define the probability specification $S^{\downarrow}$ of $v^{\downarrow}$. In the specification $S^{\downarrow}$ we need to provide a value for $p(v_i|par_i^{G^{\downarrow}})$ for each value v_i of V_i and assignment $par_i^{G^{\downarrow}}$ of the parents $Par_i^{G^{\downarrow}}$ of V_i in $G^{\downarrow}$. If V_i only occurs once in the recursive Bayesian network determined by v then we can define

$$p(v_i|par_i^{G^{\downarrow}}) = p(v_i|dsup_i^v \wedge par_i^v) = p_{dsup_i^v}(v_i|par_i^v),$$

which is provided in the specification of the value of V_i's direct superior in v. If V_i occurs more than once in the recursive Bayesian network determined by v then the specifications of v contain $p_{dsup^G}(v_i|par_i^G)$ for each graph G in v in which V_i occurs. Then $DSup_i^v = \bigcup_G DSup_i^G$ and $Par_i^v = \bigcup_G Par_i^G$, with the unions taken over all such G. Now the specifiers $p_{dsup^G}(v_i|par_i^G)$ constrain the value of

---

[8]Contraction is the following property of probabilistic independence: $R \perp\!\!\!\perp S|T$ and $R \perp\!\!\!\perp U|S, T \Rightarrow R \perp\!\!\!\perp S, U|T$. See e.g. [Pearl, 1988].

$p_{dsup_i^v}(v_i|par_i^v)$ but may not determine it completely. These are linear constraints, though, and thus there is a unique value for $p_{dsup_i^v}(v_i|par_i^v)$ which maximises entropy subject to the constraints holding—this can be taken as its optimal value, [9] and $p(v_i|par_i^{G^\downarrow})$ can be set to this value. [10]

Having fully defined the flattening $v^\downarrow = (G^\downarrow, S^\downarrow)$ and shown that the causal Markov condition holds, we have a (non-recursive) Bayesian network, [11] which can be used to determine a probability function over assignments to v:

PROPOSITION 5. *A recursive Bayesian network determines a unique joint distribution over consistent acyclic assignments v of values to its domain, defined by*

$$p(v) = \prod_{i=1}^n p(v_i|par_i^{G^\downarrow})$$

where $G^\downarrow$ is the graph in the flattening $v^\downarrow$ of v and $p(v_i|par_i^{G^\downarrow})$ is the value in the specification $S^\downarrow$ of $v^\downarrow$. (As usual v_i is the value v assigns to V_i and $par_i^{G^\downarrow}$ is the assignment v gives to the parents of V_i according to $G^\downarrow$.) [12]

While a flattening is a useful concept to explain how a joint distribution is defined, there is no need to actually construct flattenings when performing calculations with recursive networks—indeed that would be most undesirable, given that there are exponentially many assignments and thus exponentially many flattenings which would need to be constructed and stored. By Proposition 5, only the probabilities $p(v_i|par_i^v \wedge dsup_i^v)$ need to be determined, and in many cases (i.e. when V_i occurs only once in v) these are already stored in the recursive network.

The concept of flattening, in which a mapping is created between a recursive network and a corresponding non-recursive network, also helps us understand how standard inference algorithms for non-recursive Bayesian networks can be directly applied to recursive networks. For example, message-passing propagation algorithms[13] can be directly applied to recursive networks, as long as messages are passed between direct superior and direct inferior as well as between parent and child. Moreover, recursive Bayesian networks can be used to reason about interventions just as can non-recursive networks: when one intervenes to fix the value of a variable one must treat that variable as a root node in the network, ignoring any

---

[9][Jaynes, 1957].

[10]See [Williamson, 2002] for more on maximising entropy.

[11]Note that this Bayesian network is not causally interpreted, since arrows from superiors to direct inferiors are not causal arrows.

[12]Here the domain of p is the set of assignments to V, and p is unique over consistent acyclic assignments. If one wants to take just the set of consistent acyclic assignments as domain of p (equivalently, to award probability 0 to inconsistent or cyclic assignments) then one must renormalise, i.e. divide $p(v)$ by $\sum p(v)$ where the sum is taken over all consistent acyclic assignments.

[13]See [Pearl, 1988], [Neapolitan, 1990].

connections between the node and its parents or direct superiors. [14] In effect, tools for handling non-recursive Bayesian networks can be easily mapped to recursive networks.

A word on the plausibility of the recursive Markov condition. It was shown in [Williamson, 2001] that the causal Markov condition can be justified as follows: suppose an agent's background knowledge consists of the components of a causally interpreted Bayesian network—knowledge of causal relationships embodied by the causal graph and knowledge of probabilities encapsulated in the corresponding probability specification—then the agent's degrees of belief ought to satisfy the causal Markov condition. [15] This justification rests on the acceptance of the maximum entropy principle (which says that an agent's belief function should be the probability function, out of all those that satisfy the constraints imposed by background knowledge, that has maximum entropy) and the causal irrelevance principle (which says that if an agent learns of the existence of new variables which are not causes of any of the old variables, then her degrees of belief concerning the old variables should not change). An analogous justification can be provided for the recursive Markov condition. Plausibly, learning of new variables that are not superiors (or causes) of old variables should not lead to any change in degrees of belief over the old domain. Now if an agent's background knowledge takes the form of the components of a recursive Bayesian network then the maximum entropy function, and thus the agent's degrees of belief, will satisfy the recursive Markov condition as well as the causal Markov condition. Thus a justification can be given for both the causal Markov condition and the recursive Markov condition.

6   Related Work

Bayesian networks have been extended in a variety of ways, and some of these are loosely connected with the recursive Bayesian networks introduced above.

Recursive Bayesian multinets generalise Bayesian networks along the following lines. [16] First, Bayesian networks are generalised to *Bayesian multinets* which represent context-specific independence relationships by a set of Bayesian networks, each of which represents the conditional independencies which operate in a fixed context. By creating a variable C whose assignments yield different contexts, a Bayesian multinet may be represented by decision tree whose root is C and whose leaves are the Bayesian networks. The idea behind recursive Bayesian multinets is to extend the depth of such decision trees. Leaf nodes are still Bayesian networks, but there may be several decision nodes. For example, Figure 24 depicts a recursive Bayesian multinet in which there are three decision nodes, C_1, C_2 and C_3, and four Bayesian networks B_1, B_2, B_3, B_4. Node C_1 has two possible con-

---

[14] [Pearl, 2000, Section 1.3.1].

[15] See also [Williamson, 2002].

[16] [Peña *et al.*, 2002].

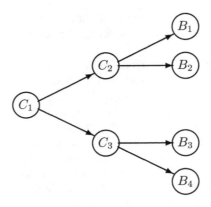

Figure 24. A recursive Bayesian multinet

texts as values; under the first node C_2 comes into operation; this has two possible contexts as values; under the first Bayesian network B_1 describes the domain; under the second B_2 applies, and so on. Figure 24 is recursive in the sense that depending on the value of C_1, a different multinet is brought into play—the multinet on C_2, B_1, B_2 or that on C_3, B_3, B_4. Thus recursive Bayesian multinets are rather different to our recursive Bayesian networks: they are applicable to context-specific causality where the contexts need to be described by multiple variables, [17] not to general instances of recursive causality, and consequently they are structurally different, being decision trees whose leaves are Bayesian networks rather than Bayesian networks whose nodes take Bayesian networks as values.

Recursive relational Bayesian networks generalise the expressive power of the domain over which Bayesian networks are defined. [18] Bayesian networks are essentially propositional in the sense that they are defined on variables, and the assignment of a value to a variable can be thought of as a proposition which is true if the assignment holds and false otherwise. We have made this explicit by representing the two possible assignments to variable A by a and $\neg a$ respectively. *Relational Bayesian networks* generalise Bayesian networks by enabling them to represent probability distributions over more fine-grained linguistic structures, in particular certain sub-languages of first-order logical languages. Recursive relational Bayesian networks generalise further by allowing more complex probabilistic constraints to operate, and by allowing the probability of an atom that instantiates a node to depend recursively on other instantiations as well as the node's

---

[17]The particular application that motivated their introduction was data clustering—see [Peña *et al.*, 2002].

[18][Jaeger, 2002].

parents.[19] Thus in the transition from relational Bayesian networks to recursive relational Bayesian networks the Markovian property of a node being dependent just on its parents (not further non-descendants) is lost. Therefore recursive relational Bayesian networks and recursive Bayesian networks differ fundamentally with respect to both motivating applications and formal properties.

Object-oriented Bayesian networks were developed as a formalism for representing large-scale Bayesian networks efficiently.[20] Object-oriented Bayesian networks are defined over *objects*, of which a variable is but one example. Such networks are in principle very general, and recursive Bayesian networks are instances of object-oriented Bayesian networks in as much as recursive Bayesian networks can be formulated as objects in the object-oriented programming sense. Moreover in practice object-oriented Bayesian networks often look much like recursive Bayesian networks, in that such a network may contain several Bayesian networks as nodes, each of which contains further Bayesian networks as nodes and so on.[21] However, there is an important difference between the semantics of such object-oriented Bayesian networks and that of recursive Bayesian networks, and this difference is dictated by their motivating applications. Object-oriented Bayesian networks tend to be used to organise information contained in several Bayesian networks: each such Bayesian network is viewed as a single object node in order to hide much of its information that is not relevant to computations being carried out in the containing network. Hence when there is an arrow from one Bayesian network B_1 to another B_2 in the containing network, this arrow hides a number of arrows from output variables (which are often leaf variables) of B_1 to input variables (often root variables) of B_2. So by expanding each Bayesian network node, an object-oriented Bayesian network can be expanded into one single non-recursive, non-object-oriented Bayesian network. In contrast, in a recursive Bayesian network, recursive indexBayesian network!recursiveBayesian networks occur as *values* of nodes not as nodes themselves, and when one recursive Bayesian network b_1 causes another b_2 in a containing recursive Bayesian network b, it is not output variables of b_1 that cause input variables of b_2, it is b_1 *as a whole* that causes b_2 *as a whole*. Correspondingly, there is no straightforward mapping of a recursive Bayesian network on V to a Bayesian network on V: mappings (flattenings) are relative to assignment v to V. Thus while object-oriented Bayesian networks are in principle very general, in practice they are often used to represent very large Bayesian networks more compactly by reducing sub-networks into single nodes. In such cases the arrows between nodes in an object-oriented Bayesian network are interpreted very differently to arrows between nodes in a recursive Bayesian network, and issues such as causal, Markov and probabilistic

---

[19] See [Jaeger, 2002] for the details.
[20] [Koller & Pfeffer, 1997].
[21] See [Neil *et al.*, 2000] for example.

consistency do not arise in the former formalism.

Hierarchical Bayesian networks (HBNs) were developed as a way to allow nodes in a Bayesian network to contain arbitrary lower-level structure.[22] Thus recursive Bayesian networks can be viewed as one kind of HBN, in which lower-level structures are of the same type as higher-level structures, namely Bayesian network structures. In fact, HBNs were developed along quite similar lines to recursive Bayesian networks, and even have a concept of flattening. However, there are a number of important differences. As mentioned, HBNs are rather more general in that they allow arbitrary structure. It is questionable whether this extra generality can be motivated by causal considerations: certainly HBNs seem to have been developed in order to achieve extra generality, while recursive Bayesian networks were created in order to model an important class of causal claims. HBNs have been developed in most detail in the case considered in this paper, namely where lower-level structure corresponds to causal connections. However, the lower-level structures are not exactly Bayesian networks in HBNs: one must specify the probability of each variable conditional on its parents in its local graph *and all variables higher up the hierarchy*. Thus HBNs have much larger size complexity than recursive Bayesian networks. HBNs do not adopt our recursive Markov condition—they only assume that a variable is probabilistically independent of all nodes that are not its descendants conditional on its parents and *all higher-level variables*. This has its advantages and its disadvantages: on the one hand it is a weaker assumption and thus less open to question, on the other it leads to the larger size of HBNs. Finally, variables can only appear once in a HBN, but they can appear more than once in a recursive Bayesian network—we would argue that repeated variables are well-motivated in terms of recursive causality (Section 1). Thus HBNs are more restrictive than recursive Bayesian networks in one respect, and more general in another, and have quite different probabilistic structure. However, they share common ground too, and where one formalism is inappropriate, the other might well be applicable.

7 Structural Equation Models

Of course, a Bayesian network is not the only type of causal model, and the extension of Bayesian networks to recursive Bayesian networks can be paralleled in other types of causal model.

After Bayesian networks, perhaps the most widely applied type of causal model is the *structural equation model*. This consists of a 'pseudo-deterministic' equation determining the value of each effect as a function of the values of its direct causes and an error variable:

$$V_i = f(Par_i, \varepsilon_i),$$

---

[22][Gyftodimos & Flach, 2002].

for $i = 1, \ldots, n$ and where Par_i is the set of direct causes of V_i and each error variable ε_i is independently distributed. Typically the function f will be linear. The effect is always written on the left-hand side of the equation; by adopting this convention one can determine the causal structure (in the shape of a causal graph) from the set of equations. Structural equations are quite restrictive—they only allow variables to vary with their direct causes (and independent error variables)— and one can prove that the causal Markov condition holds given this restriction. If we specify the probability distribution of each root variable (the variables which have no causes) then we have a Bayesian network, since a structural equation determines the probability distribution of each non-root variable conditional on its parents in the causal graph. A Bayesian network does not determine pseudo-deterministic functional relationships however, and so a structural equation model is a stronger kind of causal model than a Bayesian network.

Structural equation models can be extended to model recursive causality as follows. A *recursive structural equation model* takes not only simple variables as members of its domain, but also *SEM-variables* which take structural equation models as values (including a level 0 variable which takes as its only value the top-level model).[23] As with recursive Bayesian networks we can impose natural consistency conditions on a recursive structural equation model: causal consistency and consistency of functional equations. Given an assignment to the domain, we can create a corresponding, non-recursive structural equation model, its *flattening*, and define a pseudo-deterministic functional model over the whole domain by constructing an equation for each variable as a function of its direct superiors as well as its direct causes (and an error variable).

We see, then, that the move from an ordinary Bayesian network to a recursive Bayesian network can be mirrored in other types of causal model. In the following sections we will study this move from a more general point of view. We will see that the strategy of rendering a general network structure recursive can be applied in various interesting ways—not just to recursive causality.

8   Argumentation Networks

Recursive networks are not just useful for reasoning with causal relationships— they can also be used to reason with other relationships that behave analogously to causality. In this section we shall briefly consider the relation of support between arguments.

In an *argumentation framework*, one considers arguments as relata and attacking as a relation between arguments.[24] Consider the following example.[25] Hal is a

---

[23] Warning: in the past, acyclic structural equation models have occasionally been called 'recursive structural equation models'—clearly 'recursive' is being used in a different sense here.

[24] [Dung, 1995].

[25] Due to [Coleman, 1992] and discussed in [Bench-Capon, 2003, Section 7].

diabetic who loses his insulin; he proceeds to the house of another diabetic, Carla, enters the house and uses some of her insulin. Was Hal justified? The argument (A_1) 'Hal was justified since his life being in danger allowed warranted his drastic measures' is attacked by (A_2) 'it is wrong to break in to another's property' which is in turn attacked by (A_3) 'Hal's subsequently compensating Carla warrants the intrusion'. This argument framework is typically represented by the picture of Figure 25.[26]

Figure 25. Hal-Carla argumentation framework

One can represent the interplay of arguments at a more fine-grained level by (i) considering propositions as the primary objects of interest, and (ii) taking into account the notion of support as well as that of attack. By taking propositions as nodes and including an arrow from one proposition to another if the former supports or attacks the latter, we can represent an argument graphically. In our example, let C represent Hal compensates Carla', B 'Hal breaks in to Carla's House', W 'Breaking in to a house is wrong' and D 'Hal's life is in danger'. Then we can represent the argument by $[C \longrightarrow^+ B] \longrightarrow^- [W \longrightarrow^- B] \longrightarrow^- [D \longrightarrow^+ B]$ (here a plus indicates support and a minus indicates attack). In general the fine structure of an argument is most naturally represented recursively as a network of arguments and propositions. We call this kind of representation a *recursive argumentation network.*

If a quantitative representation is required, recursive Bayesian networks can be directly applied here. The nodes or variables in the network are either *simple arguments*, i.e. propositions, taking values true or false, or *network arguments*, which take recursive Bayesian networks as values. In our example C is a simple argument with values c or $\neg c$ while A_2 is a network argument with values a_2 referring to $W \longrightarrow B$ (with associated probability specifiers $p(w), p(b|\pm w)$) or $\neg a_2$ representing W, B (with $p(w), p(b)$). Instead of interpreting the arrows as causal relationships, indicating causation or prevention, we interpret them as support relationships, indicating support or attack. The probability $p(v_i|par_i)$ of an assignment v_i to a variable conditional on an assignment par_i to its parents is interpreted as the probability that v_i is *acceptable* given that par_i is *acceptable*. Thus instead of representing support or attack by pluses and minuses, degree of support is represented by conditional probability distributions. If consistency and acyclicity conditions are satisfied, non-local degrees of support can be gleaned from the joint probability distribution defined over all variables.

---

[26][Bench-Capon, 2003].

Note that Bench-Capon argues that the evaluation of an argument may depend on accepted values.[27] In our example, the evaluation of the argument depends on whether health is valued more than property, in which case property argument A_2 may not defeat health argument A_1, or vice versa. These value propositions can be modelled explicitly in the network, so that, for example, A_1 depends on value proposition 'health is valued over property' as well as argument A_2.

In sum, relations of support behave analogously to causal relations and arguments are recursive structures; these two observations motivate the use of recursive Bayesian networks to model arguments. This leads us in turn to the question of how to characterise the concept of an abstract recursive network. In the next two sections we explore this question in the context of *input-output* or *information networks*.

9   Self-Fibring Networks: Overview

This section shows how our recursive network approach fits within a more general concept of substituting one network inside another (referred to as self-fibring of networks).

We will focus attention on *information networks*, which are directed acyclic graphs whose roots are *inputs*, whose leaves are *outputs* and whose arrows indicate the flow of information from input to output. Thus if we have

then we propagate the input from t_i into t. If $V(x)$ is the value at node x, then we need a propagation function f yielding $V(t) = f(V(t_1), \ldots, V(t_n))$. Note that there may be a constraint $\phi(t_1, \ldots, t_n)$ on the inputs: only if a set of value of inputs satisfies ϕ will those values be admissible.

In Bayesian networks, the arrows correspond to causal direction rather than to the flow of information. But an *application* of a Bayesian network can be construed as an information network as follows. When a Bayesian network is applied, the values of a set of variables are observed. These variables are the inputs. They are instantiated to their observed values in the Bayesian network, and this change is propagated around the network, typically using message-passing algorithms,[28]

---

[27][Bench-Capon, 2003, Section 5].
[28][Pearl, 1988].

until the probabilities of further variables of interest (the outputs) can be ascertained. Thus in message-passing algorithms information flows from the inputs to the outputs via the arrows of the Bayesian network, though not normally in accordance with the direction of the arrows in the original Bayesian network. Suppose for instance that Figure 26 is the graph of a Bayesian network, that the value of B is observed and that the probability of C is required. Then in determining the probability of C, information flows from B to C along the pathways between B and C of the original Bayesian network graph, as depicted in Figure 27.

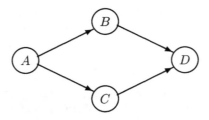

Figure 26. The graph of a Bayesian network.

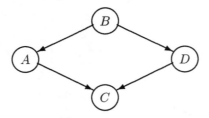

Figure 27. The graph of a corresponding information network.

Note that in general the information network is only a schematic representation of the flow of information: in fact in Bayesian network message-passing propagation algorithms, messages are passed in both directions along arrows, two passes are made of the network, and in multiply connected graphs such as Figure 26 propagation takes place in an associated undirected tree-shaped Markov network formed from the Bayesian network.[29] In singly-connected Bayesian networks though there is a fairly close correspondence between information network and flow of messages.

The question now arises as to how information networks can be *self-fibred*, i.e. substituted one inside the other.

---

[29]See [Lauritzen & Spiegelhalter, 1988].

There are several options for self fibring. We explain them briefly here and the full definitions come in the next section.

Let $\boldsymbol{B}(X)$ be a network with node X in it. Let $\boldsymbol{A}$ be another network. We want to define $\boldsymbol{C} = \boldsymbol{B}(X/\boldsymbol{A})$, a new network which is the result of substituting $\boldsymbol{A}$ for X.

Already at this stage there are several views to take.

View 1: Syntactical Substitution

Regard the operation at the syntactical level. Define $\boldsymbol{C}$ syntactically and give it meaning / semantics / probabilities derived from the meanings of $\boldsymbol{B}$ and $\boldsymbol{A}$.

View 2: Semantic Insertion

Look at the meaning of $\boldsymbol{B}$ and then define what $\boldsymbol{B}(X/\boldsymbol{A})$ is supposed to be. Here the substitution is not purely syntactic. For example, if $\boldsymbol{B}$ is a Bayesian network where the node X can take two values 0, 1 then if X is 0 we substitute (in a certain way) $\boldsymbol{A}_0$ for X and if X is 1 we substitute $\boldsymbol{A}_1$ (this is the approach taken in Section 3). The "substitution" need not be actual substitution but some operation $Ins(X, \boldsymbol{B}, \boldsymbol{A})$ inserting $\boldsymbol{A}_i$ at the point X inside $\boldsymbol{B}$.

So for example in logic we can have

$$Ins(X, X \to B, C) \stackrel{\text{def}}{=} (X \to C) \to B.$$

Thus in this case

$$Ins(X, \boldsymbol{B}(X), C) = \boldsymbol{B}(X/X \to C).$$

More complex insertions are possible for Bayesian nets. We could convert in the above case the semantic inversion into a syntactic one by splitting each variable Y in the net into two variables Y_0 (for $Y = 0$) and Y_1 (for $Y = 1$). We discuss such manipulations in the next section.

To study our options and to illustrate the ideas of self fibring we begin with a simple two point network

The input gives value to A and this is propagated to B, using the function f.

We now give several interpretations for this as implication.

Interpretation 1

The above represents a substructural implication $A \to B$. The semantical interpretation for the substructural $\to$ is via evaluation into an algebraic semigroup $(S, \circ, e)$, where $\circ$ is a binary associative operation and e is the identity.

If the wff $A \to B$ gets value t and the input A get a value a then B gets value $b = t \circ a$.

Here the network function f can be taken as the function

$$\lambda x f_t(x) = \lambda x (t \circ x).$$

Interpretation 2

This interpretation is the modus ponens in a Labelled Deductive System. The rule has the form

$$\frac{\alpha : A, \ \beta : A \to B, \ \varphi(\beta, \alpha)}{f(\beta, \alpha) : B}.$$

Its meaning is that if we prove A with label α and $A \to B$ with label β and (β, α) satisfy the enabling condition φ, then we can deduce B with label $f(\beta, \alpha)$.

The Dempster-Shafer rule is a special case of this. The Dempster-Shafer set up allows for certainty values for $A, B, A \to B$, to be closed intervals of real numbers. Thus if A has value in the real closed interval $[a, b]$ and the implication $A \to B$ has value in the interval $[c, d]$, then B has value in the interval

$$[a, b] \circ [c, d] = \left[\frac{ad + bc - ac}{1 - k}, \frac{bd}{1 - k} \right]$$

with $k = a(1 - d) + c(1 - b)$.

The side condition φ is $\varphi([a, b], [c, d])$ is that $k \neq 1$. Thus to interpret the labelled implication in our network we need to add φ to the link.

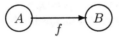

and we have:

$$\lambda x f_\beta(x) = \lambda x f(\beta, x).$$

Interpretation 3

Intuitionistic formulas as types. $A, B, A \to B$ are understood as λ calculus types having λ terms inhabiting them. We read

as a network, which for any term t of type A, given as input, the network outputs the $f(t)$ term of type B.

Thus f is of type $A \to B$.

Interpretation 4

We can regard

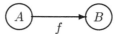

as a causal Bayesian network. The variable A can take states $a_1, \ldots, a_k$ and the variable B can take states $b_1, \ldots, b_m$ then the table f must give the conditional probability $P(B|A)$, giving the probability p_{ij} of B being in the state b_j, given A is in the state a_i. We must have $\sum_j p_{ij} = 1$.

The matrix is

$$P = \begin{pmatrix} p_{11} & p_{1m} \\ \vdots & \vdots \\ p_{k1} & p_{km} \end{pmatrix}$$

If q_i is the probability of A being in the state a_i, $(\Sigma q_i = 1)$ then the probability of B being in the state b_j is $r_j = \sum_i p_{ij} q_i$.

The logical interpretations (1)–(3) allow us to give meaning to self fibred networks where we substitute a network within a network. We have the options here of syntactical substitution (view 1) or semantical insertion (view 2). Our paper chooses semantical insertion, where we have one insertion for $X = 1$ and another for $X = 0$. In the flattening, the insertion makes X a parent to all nodes in the substituted network.

The diagram below shows how this works for $\boldsymbol{B}(X) = X \to C$ and $\boldsymbol{A}_0 = A \to B$.

For the case $X = 0$ we get the network

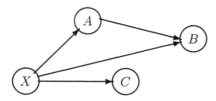

and for the case $X = 1$, we get

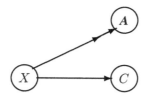

where $X \twoheadrightarrow A$ means that X connects directly to all elements in A.

We can adopt a view closer to that of logic and have no insertion if $X = 0$ and yes insertion in case $X = 1$ of a single network. The simplest cases are the following:

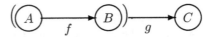

and

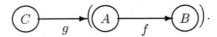

In the first case we took the network

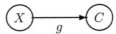

and substituted for X the network

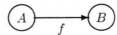

The second case is similar.

The question is what meaning do we give to these fibred networks?

Let us consider the first case

The machinery of

is to accept inputs x of a certain kind at the node X and output $g(x)$ at node C. By letting $X = A \to B$ we must ask: What is the input we are getting for X? We can say the obvious answer is that the input is f. We now have to check whether f is of the kind that can be accepted in our network.

Let us check.

Interpretation 1

Here f can be identified with an element t of the semigroup. So it is OK.

Interpretation 2

Here f can be identified with a label. So again it is OK.

Interpretation 3

Here f is a λ-term of type $A \to B$. To be OK we say X (accepts elements) of type $A \to B$ and g is of type $(A \to B) \to C$. So again we are OK.

Interpretation 4

Here A is a probability distribution for the states of A and f is a matrix of conditional probabilities. We have to deal with that! First let us simplify and say both A, B are two state variables $A = 0, A = 1, B = 0, B = 1$. Even with this simplification, still f is a 2×2 matrix P. It allows for many states not just two. This is not exactly right.

What are our options?

Option 1

Allow for new kinds of inputs for our variables. This option is complicated because of repeated iteration of fibring. We will not pursue it.

Option 2

Extract from the new input (the matrix) a recognisable input for X in $X \to C$, (i.e. a two state input). This method is what we usually do in the area of fibring logics. We need a fibring function F that will extract two states, yes or no, out of the matrix P. The function is as follows;

- yes if B depends on A in any way

- no if not.

In other words, we read X as a variable getting 1 if the network substituted for it is "on" or "active" and 0 if it is not on.

So for example if the matrix is

$$\begin{pmatrix} p & 1-p \\ p & 1-p \end{pmatrix}$$

we get

$$(q, 1-q) \begin{pmatrix} p & 1-p \\ p & 1-p \end{pmatrix} = (p, 1-p)$$

and thus the probability of B is independent of that of A.

We can talk about the probability of B being independent of A, etc.

Say we have for $0 \leq \varepsilon \leq 1 - p$

$$(q, 1-q) \begin{pmatrix} p & 1-p \\ p+\varepsilon & 1-p-\varepsilon \end{pmatrix} =$$

(2)

$$(pq + (1-q)p + (1-q)\varepsilon, (1-p)q + (1-q)(1-p) - (1-q)\varepsilon) =$$

$$(p + (1-q)\varepsilon, (1-p) - (1-q)\varepsilon).$$

The variation is $2\varepsilon(1-q) \leq 2\varepsilon$. So we can give a probability for $\varepsilon = 0$ or $\varepsilon \neq 0$.

We leave this aspect for a moment and discuss the other possibility of fibring, namely

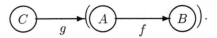

Here we substitute a network

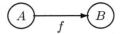

for the variable Y in

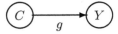

The first three interpretations will cope with this very well, because the output of g can modify the f, as they are of the same kind. Can we do something similar in the probabilities case? We again have several options:

1. We can read Y as a variable getting $0, 1$ values indicating whether the network

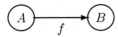

is *on* or not. The value of Y is obtained in the network

2. The second option is to use the network up to Y to modify the network which we substitute for Y.[30]

We note that g is a matrix and so is f. Should we modify f by multiplying it by g and set something like

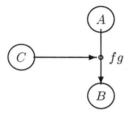

How would this relate to the network:

Let us check

has the matrix

$$\begin{pmatrix} p_1, & 1 - p_1 \\ p_2, & 1 - p_2 \end{pmatrix}$$

and

---

[30] In case of neural networks this is the more reasonable option.

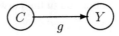

has the matrix

$$\begin{pmatrix} \gamma_1, & 1-\gamma_1 \\ \gamma_2, & 1-\gamma_2 \end{pmatrix}$$

gf is

$$\begin{pmatrix} \gamma_1 & 1-\gamma_1 \\ \gamma_2 & 1-\gamma_2 \end{pmatrix} \cdot \begin{pmatrix} P_1 & 1-P_1 \\ P_2 & 1-P_2 \end{pmatrix} =$$

(3)

$$\begin{pmatrix} \gamma_1 p_1 + (1-\gamma_1)p_2, & \gamma_1(1-p_1) + (1-\gamma_1)(1-p_2) \\ \gamma_2 1 p_1 + (1-\gamma_2)p_2, & \gamma_2(1-p_1) + (1-\gamma_2)(1-p_2) \end{pmatrix}$$

This would interpret

as

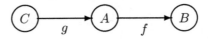

We prefer the first option.

Let us now see what to do with networks of the form

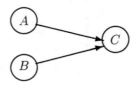

Can we read the $\rightarrow$ as implication? The answer is yes for the first three "logical" interpretations. We read it as

$$\langle A, B \rangle \rightarrow C$$

or

$$A \otimes B \rightarrow C$$

where $\otimes$ is a commutative binary operation. It is the multiplicative conjunction in linear logic and is the ordinary conjunction in intuitionistic logic. We have in case of logic that:[31]

$$(A \otimes B \rightarrow C) \equiv (A \rightarrow (B \rightarrow C)) \equiv (B \rightarrow (A \rightarrow C))$$

This does not hold in the Bayesian network case. As defined in Section 2 we need a function giving a probability value for C, for each pair of possible values (x, y) for A, B.

We still need to give meaning to

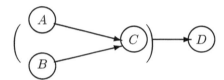

and

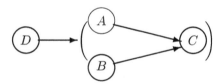

The first is obtained by substituting the network

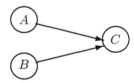

---

[31] For the Dempster–Shafer rule we calculate $[a, b] \otimes [c, d]$ as $[a, b] \circ [c, d]$.

for X in

and the other in

The principles we discovered still hold. In the first case the fibring function gives X values $0, 1$ depending whether we believe in the connection between A, B, C. i.e. the network is "on" or not.

The second case would require modifying the network of A, B, C by using the network

The simplest is to take option 1. The value $x = 1$ means the network with A, B, C is "on" and otherwise it is not.

In any case, the kind of choices we have to make are clear! There is a lot of scope for fine tuning. For example we can look at

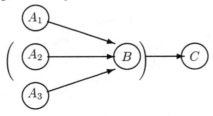

as a family of networks of the form

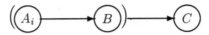

using the probabilities in the substituted network (fix A_j, $j \neq i$ as 0,1) to decide on priorities.

10   Self-Fibring Networks: Theory

The aim of this section is to present a general theory of networks and fibring of networks, in order to put our work into perspective. We begin with an example. Figure 28 shows a typical network.

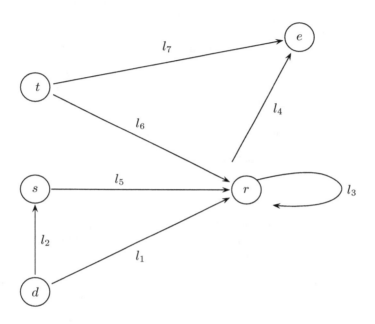

Figure 28.

The nodes of the network is the set $S = \{d, s, t, r, e\}$. We consider d as the input point and e as the output point. The arrows represent connections between nodes. This is a binary relation $R \subseteq S^2$. In Figure 28 we have $R = \{(d, r), (s, r), (t, r), (t, e), (r, e), (r, r), (d, s)\}$. The labels l decorate the connections. So let τ be a function from R into a set of labels L. In Figure 28 we have $L = \{l_1, \ldots, l_7\}$ and $\tau((d, r)) = l_1, \ldots, \tau((t, e)) = l_7$. In addition the nodes are coloured by a colouring function (values which we call colours) giving values in some space Ω. Thus $V(t), V(d), \ldots$ are the colours of the nodes. In Bayesian nets for example, $V(t)$ is a probability.

We also require a family of propagation functions $\mathbb{F}$, giving values in the space

Ω of the form below and such that the following recursive equation holds:

$$V(t) = \mathbb{F}(t, (\tau_1, V(x_1)), (\tau_2, V(x_2)), \ldots, (\tau_n, V(x_n)))$$

where $x_1, \ldots, x_n$ are all the parents of t (i.e. all x_i such that $(x_i, t) \in R$), τ_i is the label of (x_i, t) and $V(x_i)$ are the colours of the nodes x_i, respectively. Note that $\mathbb{F}$ can operate on any number of variables, i.e. n can be arbitrary.

We perceive the colouring to propagate along the network using the arrows, the labels and the function $\mathbb{F}$. In case the network has cycles, we expect V to be implicitly defined by $\mathbb{F}$.

The network in Figure 28 can be interpreted in several ways:

1. It can be interpreted as a map, where the nodes are towns, the labels are distances and the colours are some heuristic numbers to aid some search function(e.g. the labels can give the aerial distance from a central point). We may require the graph to be acyclic. The function $\mathbb{F}$ can give the average distance of the parent nodes from the current node.

2. The network can be Bayesian, in which case we require it to be acyclic. We also require any point $t \neq d$ to either have a parent $\neq d$ (i.e. for some x, $(x, t) \in R, x \neq d$) or to have d alone as a parent. We forbid d itself to have parents.

   Thus d is a dummy point $(d = \top)$ showing the nodes without parents in the rest of the network. The function $\mathbb{F}$ would be the conditional probabilities of a node on its parents.

3. The network can be a neural net with τ, V different weights on the nodes and connections, and $\mathbb{F}$ some meaningful averaging function.

4. The network can be describing a flow problem with τ giving capacities, V giving retention and $\mathbb{F}$ is the obvious function summing up the flow.

We now give a formal definition of a network.

DEFINITION 6. A network has the form

$$\mathcal{N} = (S, R, d, e, \tau, V, L, \mathbb{F}, \Omega)$$

where S is the set of nodes. $d, e \in S$ are the input and output nodes. (We may have several i.e. d_i, e_j.) τ is a labelling function $\tau : R \mapsto L$. L is a set of labels, V is a colouring function on S (range of V is in Ω) and $\mathbb{F}$ is a function giving some value in the space Ω to any finite list of the form $(t, (V_1, l_1), \ldots, (V_n, l_n))$ and the following is required to hold for any $t \in S$ and x_i such that $(x_i, t) \in R$.

- $V(t) = \mathbb{F}(t, (V(x_1), \tau((x_1, t))), \ldots, (V(x_n), \tau(x_n, t))))$ where x_i are *all* the parents of t.

DEFINITION 7 (Fibring function). Let $\mathbb{F}$ be a propagation family for states S and labels L with values in the space Ω. Then a function $\mathbf{F}$ giving a new propagation function for any triple $(V, l_i, \mathbb{F}), i = 1, \ldots, n$ is called a fibring function. We write $\mathbf{F}$ as

$$\mathbf{F} : (V, l_i, \mathbb{F}) \mapsto \mathbb{F}_{V, l_i}.$$

so $\mathbf{F}$ is defined for any set of labels.

Note that n is variable.

We now need to make some distinctions about fibring of network within networks. We give some additional examples.

EXAMPLE 8 (Refinement). Consider a network which is a map

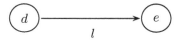

Figure 29.

So ⓓ may be Durham and ⓔ is Edinburgh. Say the label l is the number of heavy trucks per day one can push through from ⓓ to ⓔ. We can try and define this map by putting in for ⓔ another network, say E which is the map on Edinburgh. This is substituting the actual sorting networks in the UK.

A third simpler example is when ⓓ and ⓔ are days and we can refine them into hours, see Figure 30

Figure 30.

EXAMPLE 9 (Cut rule in logic). We get fibring/substitution of networks when we consider versions of the cut rule in Labelled Deductive Systems. We give a simple case. Assume our data is a list of formulas, and our language contains $\rightarrow$ only. Thus for example, we may have the list of Figure 31. We can perform modus ponens between any $X \rightarrow Y$ and X, provided X is immediately to its right and the result Y *replaces* $(X \rightarrow Y, X)$ in the list. This way of doing modus ponens

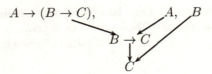

Figure 31.

characterises the one arrow *Lambek Calculus*. How would cut work? Suppose we have a proof of A in Figure 32

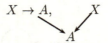

Figure 32.

We can simply substitute the sequence or net for A to get Figure 33

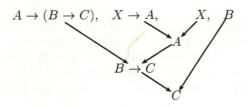

Figure 33.

Suppose now that $X \to Y$ means a version of strict implication.

- If X holds next day then Y holds next day.

The sequence
$$A \to (B \to C), A, B$$
can still be reduced to C but we must keep count of the days.
    Consider
$$C \to E, A \to (B \to C), A, B$$
We can get Figure 34 in the Lambek Calculus.

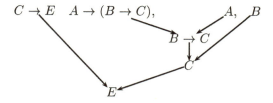

Figure 34.

This will not work in the modal strict implication meaning of → because we must follow Figure 35:

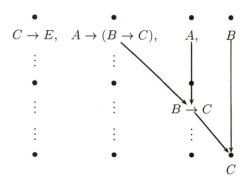

Figure 35.

$C \to E$ is 3 days away from C. We need something like $\top \to (\top \to (C \to E))$.

Thus the network substitution of $(X \to A, X)$ into $(A \to (B \to C), A, B)$ should be different in the strict implication case.

It should give the result in Figure 36

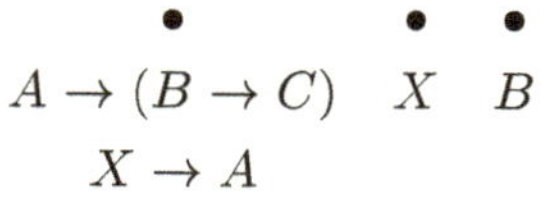

$$A \to (B \to C) \quad X \quad B$$
$$X \to A$$

Figure 36.

To summarise: $(X \to Y, X)$ is replaced by (Y) in the Lambek Calculus and is replaced by $(\cdot, Y)$ in the strict implication logic.

Thus the network substitutions corresponding to these logics are as follows:

Let
$$N_1 = (x_1, \ldots, x_n, y, z_1, \ldots, z_m)$$
$$N_2 = (u_1, \ldots, u_k)$$

then the *Lambek substitution* is:

$$N_1(y/N_2) = (x_1, \ldots, x_n, u_1, \ldots, u_k, z_1, \ldots, z_m).$$

and the *Strict substitution* is
(when $k = n + 1$)

$$N_1(y/N_2) = (x_1 \wedge u_1, \ldots, x_n \wedge u_n, u_{n+1}, z_1, \ldots, z_m)$$

If $k \neq n$, add $\top$s to the beginning of the shorter one to make them equal and then substitute.

The moral of this example is that if the networks represent some logic, then options for fibring networks represent options for the cut rule in the logic.

REMARK 10. We need to prepare the ground for the general definition of fibring which will follow.

Assume $\mathcal{N}_1$ and $\mathcal{N}_2$ are as in Figures 37 and 38

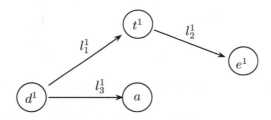

Figure 37.

We have several options in substituting $\mathcal{N}_2$ for t^1, using a fibring function **F**. The most straightforward one is to replace t^1 by $\mathcal{N}_2$ and redirect all arrows coming into t^1 and connect them to all input points d_j^2 of $\mathcal{N}_2$. Similarly all arrows coming out of t^1 will now come out of every output point e_j^2 of $\mathcal{N}_2$.

Figure 39 shows the result.

The function $\mathbb{F}^{1,2}$ is the same as $\mathbb{F}^1$ on nodes from $\mathcal{N}_1$ and is the fibred function $\mathbb{F}^2_{V_{(t^1)}^1, l_1^1, l_2^1}$ obtained by applying **F** to $\mathbb{F}^2$.

Variations can be obtained by changing $\mathbb{F}$ and/or by changing the input output points or $\mathcal{N}_2$ before fibring. So this is quite a general definition. The basic idea is

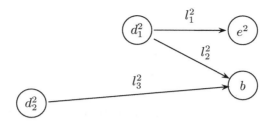

Figure 38.

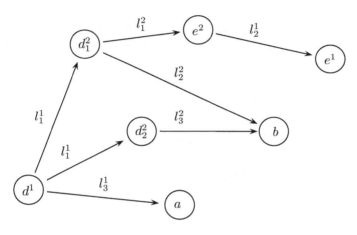

Figure 39.

that the 'environment' of t^1 (namely V^1/t^1) and all labels of connections leading into and out of t^1) change the fibring function $\mathbb{F}^2$ of the substituted network $\mathcal{N}_2$ into $\mathbf{F}(\mathbb{F}^2)$.

Problems may arise if either $\mathcal{N}_1$ and $\mathcal{N}_2$ have nodes in common or if t^1 is connected to itself. This can cause more than one arrow to occur between two points. For this reason these situations are excluded. To see why this can happen, imagine we substitute

into

where t is both the input and output points. We get by definition the network

NOTATION 11. If $\mathcal{N}_1, \mathcal{N}_2$ is described using $\rightarrow$ and t^1 is in $\mathcal{N}_1$, we can indicate fibring by substituting $\mathcal{N}_2$ for t^1 and using $\twoheadrightarrow$ to connect into and out of $\mathcal{N}_2$.
The fibring function $\mathbf{F}$ is suppressed.

Here is an example for our notation:

EXAMPLE 12. Example of network using $\rightarrow$:

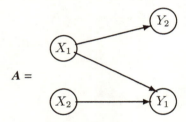

We use $\twoheadrightarrow$ as a special connection between a node X and a network A.

If A is the network above and $\twoheadrightarrow$ means that we $\rightarrow$ connect with every node in the network than $X \twoheadrightarrow A$ means in this case

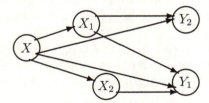

Arrows coming out of $\mathbf{A}$ into Y are not drawn.

We now conclude with a general definition:

DEFINITION 13. Let $\mathcal{N}_i = (S^i, R^i, d^i_j, e^i_k, \tau^i, V^i, L^i, \mathbb{F}^i, \Omega^i)$ for $i = 1, 2$, be several networks based on the same set S of nodes (i.e. $S^i \subseteq S$) and sets L and

Ω (i.e. $L^i \subseteq L, \Omega^i \subseteq \Omega$). Assume $S^1 \cap S^2 = \varnothing$. Let $\mathbf{F}$ be a fibring function and let $t^1 \in S^1$ be a node such that $(t^1, t^2) \notin R^1$ and let l^1_j be all the labels of nodes in $\mathcal{N}_1$ leading into or coming out of t^1. We define the one step fibred system $\mathcal{N}_{1,2} = \mathcal{N}_1(t^1 / \mathcal{N}_2)$ as follows:

1. $S^{1,2} = (S^1 \cup S^2) - \{t^1\}$.

2. $R^{1,2} = (R^1 \cup R^2 \cup \{(x, d^2_i)|(x, t^1) \in R^1\} \cup \{(e^2_j, y)|(t^1, y) \in R^1\}) - \{(x, y) \in R^1 | x = t^1 \text{ or } y = t^1\}$.

3. $\{d^{1,2}_i\} = \{d^1_i | t^1 \neq d^1_i\} \cup \{d^2_k | t^1 \text{ is an input point}\}$.

4. $\{e^{1,2}_j\} = \{e^1_j | t^1 \neq e^1_j\} \cup \{e^2_k | t^1 \text{ is an output point}\}$.

5. $\tau^{1,2}((x, y)) = \bullet \tau^1((x, y))$ if $x, y \in S^1$ and $(x, y) \in R^{1,2}$

   $\bullet \tau^2((x, y))$ if $(x, y) \in R^{1,2}$ and $x, y \in S^2$

   $\bullet \tau^1((x, t^1))$ if $x \in S^1, y \in S^2, (x, y) \in R^{1,2}$

   $\bullet \tau^1((t^1, y))$ if $x \in S^2, y \in S^1$ and $(x, y) \in R^{1,2}$

6. $\Omega^{1,2} = \Omega^1 \cup \Omega^2$

7. $L^{1,2} = L^1 \cup L^2$

8. $\mathbb{F}^{1,2}$ is defined as $\mathbf{F}(V^1(t^1), l^1_i, \mathbb{F}^2)$ where l_i are all the labels from other nodes in S^1 leading to t^1 and labels of nodes in S^1 into which t^1 leads. We assume $\mathbf{F}$ is such that $\mathbb{F}^{1,2} = \mathbb{F}^1$ on points in S^1. This is possible since we assumed $S^1 \cap S^2 = \varnothing$.

11 Conclusion

The moral of this paper is this: recursive structures are rife and benefit from explicit modelling. If causality is to be modelled then it is not always enough to rely on Bayesian networks, for these fail to model recursive structure. Recursive Bayesian networks can be used, however, and these admit joint distributions just as do non-recursive Bayesian networks. Analogously structural equation models can be extended to recursive structural equation models. Recursive Bayesian networks can also be applied to non-causal domains, such as argumentation. A very general type of recursive input-output network, called a self-fibred information network, extends these models and admits interesting applications in logic, where arrows are interpreted as implication.[32,33]

---

[32]Further results on argumentation networks and on fibring neural networks can be found in [Gabbay & Woods, 2003], [Barringer et al., 2003] and [Garcez et al., 2003].

[33]We thank David Glass for many helpful comments.

BIBLIOGRAPHY

[Barringer *et al.*, 2003] H. Barringer, D. Gabbay & J. Woods. Temporal dynamics of argumentation networks, draft.

[Bench-Capon, 2003] T.J.M. Bench-Capon. Persuasion in practical argument using value based argumentation frameworks. *Journal of Logic and Computation*, **13**, 429–448, 2003.

[Coleman, 1992] J. Coleman. *Risks and Wrongs*. Cambridge University Press, 1992.

[Corfield & Williamson, 2001] D. Corfield & J. Williamson, eds. *Foundations of Bayesianism*. Kluwer Applied Logic Series, Dordrecht: Kluwer Academic Publishers, 2001.

[Dung, 1995] Phan Minh Dung. On the acceptability of arguments and its fundamental role in nonmonotonic reasoning, logic programming and n-person games. *Artificial Intelligence*, **77**, 321–357, 1995.

[Gabbay & Woods, 2003] D. Gabbay & J. Woods. The laws of evidence and labelled deduction, *Phinews*, October, 5–46, 2003.

[Garcez *et al.*, 2003] A. Garcez, L. Lamb & D. Gabbay. Fibring of neural nets, draft.

[Gyftodimos & Flach, 2002] E. Gyftodimos & P. Flach. Hierarchical Bayesian Networks: A Probabilistic Reasoning Model for Structured Domains. In E. de Jong & T. Oates, eds. *Proceedings of the International ConferenceML-2002 Workshop on Development of Representations*, pp. 23–30. University of New South Wales, 2002.

[Jaeger, 2002] M. Jaeger. Complex probabilistic modeling with recursive relational Bayesian networks. *Annals of Mathematics and Artificial Intelligence*, to appear.

[Jaynes, 1957] E.T. Jaynes. Information theory and statistical mechanics. *The Physical Review*, **106**, 620–630, 1957.

[Koller & Pfeffer, 1997] D. Koller & A. Pfeffer. Object-oriented Bayesian networks. In *Proceedings of the Thirteenth Annual Conference on Uncertainty in Artificial Intelligence*, pp. 302–313, 1997.

[Lauritzen & Spiegelhalter, 1988] S.L. Lauritzen & D.J. Spiegelhalter. Local computation with probabilities in graphical structures and their applications to expert systems, with discussion. *Journal of the Royal Statistical Society*, **B 50**, 157–254, 1988.

[Mellor, 1995] D.H. Mellor. *The Facts of Causation*. London and New York: Routledge, 1995.

[Neal, 2000] R. M. Neal. On deducing conditional independence from d-separation in causal graphs with feedback. *Journal of Artificial Intelligence Reasearch*, **12**, 87–91, 2000.

[Neapolitan, 1990] R.E. Neapolitan. *Probabilistic reasoning in expert systems: theory and algorithms*. New York: Wiley, 1990.

[Neil *et al.*, 2000] M. Neil, Norman Fenton & Lars Neilsen. Building large-scale Bayesian networks. *The Knowledge Engineering Review*, **15**, 257–284, 2000.

[Pearl, 1988] J. Pearl. *Probabilistic Reasoning in Intelligent Systems: Networks of Plausible Inference*. San Mateo, CA: Morgan Kaufmann, 1988.

[Pearl, 2000] J.Pearl. *Causality: Models, Reasoning, and Inference*. Cambridge University Press, 2000.

[Peña *et al.*, 2002] J.M. Peña, J.A. Lozano & P. Larrañaga. Learning recursive Bayesian multinets for clustering by means of constructive induction. *Machine Learning*, **47**, 63–90, 2002.

[Spirtes, 1995] P. Spirtes. Directed cyclic graphical representation of feedback models. In *Proceedings of the 11th Conference on Uncertainty in Artificial Intelligence*, Montreal QU, pp. 491–498. Morgan Kaufmann, 1995.

[Williamson, 2001] Jon Williamson. Foundations for Bayesian networks. In [Corfield & Williamson, 2001, pp. 75–115].

[Williamson, 2002] Jon Williamson. Maximising entropy efficiently. *Electronic Transactions in Artificial Intelligence Journal*, **6**, 2002. www.etaij.org.

Jon Williamson
Department of Philosophy
King's College
Strand
London, WC2R 2LS, UK
Email: jon.williamson@kcl.ac.uk
and

Dov Gabbay
Department of Computer Science
King's College
Strand
London, WC2R 2LS, UK
Email: dg@dcs.kcl.ac.uk

Recursive Causality and the Causal Relata: Comments on Williamson and Gabbay

DAVID H. GLASS

1 Introduction

Williamson and Gabbay have presented a number of exceptionally interesting ideas on recursive causality, how it can be represented in Bayesian networks and how this work can be linked to argumentation and self-fibring networks. In addition to the main focus of their paper they have given helpful overviews of these areas of artificial intelligence and have managed to make a number of technical contributions along the way which are significant in their own right.

The fundamental idea in the paper is that of recursive causality which is the notion that causal relations can themselves be causes. This is a very interesting claim and raises a number of important questions regarding the nature of causality and the causal relata. Before discussing these issues concerning the metaphysics of causality in more detail, however, I will outline some of the other developments presented in their paper.

In Section 3 the authors discuss the extension of recursive causality to Bayesian networks. Their approach is to allow variables in a Bayesian network to take on Bayesian networks as values. A recursive Bayesian network is the name given to a network which contains such variables (network variables). Since variables in a recursive Bayesian network can occur at different levels[1] there is a danger that the causal and probabilistic relations at these levels will contradict each other unless consistency checks are carried out. How can consistency be ensured? Since the problem of determining when two non-recursive Bayesian networks are consistent has not received much attention in the literature, the authors first take on the task of addressing this problem before applying their approach to the recursive case. In order to have consistency between two Bayesian networks they point out that there must be causal consistency, Markov consistency (i.e. consistency

---

[1]Consider, for example, a network which represents restrictions placed on tobacco advertising by the government (A) caused by the fact that smoking causes cancer (SC). Level 1 contains the nodes A and SC while level 2 has variables for smoking (S) and cancer (C) and represents the causal relationship between smoking and cancer. In this example, the variable SC is referred to as a *direct superior* of S and C, which in turn are *direct inferiors* of SC.

Laws and Models in in Science, 223–232.

concerning claims of probabilistic independencies) and probabilistic consistency (i.e. consistency concerning probabilistic specifications). However, it turns out that these three types of consistency are not sufficient to ensure that two consistent Bayesian networks determine the same probability function over their shared variables. In particular, the requirement of Markov consistency, whereby the networks have the same conditional independencies on shared variables, needs to be strengthened by the condition that their shared nodes are *closed under closest common causes*.[2] This is an interesting result and shows that the issue of consistency is not straightforward even in the non-recursive case.

The discussion of consistency is then applied to recursive Bayesian networks. For an assignment to the nodes of the recursive network each pair of networks contained within it are treated non-recursively and checked for consistency. While this work on consistency is crucial for recursive networks, it is also of more general interest. The authors have given a clearly defined notion of consistency for a pair of Bayesian networks which may well be of relevance to work in the area of information fusion, where possibly conflicting information is combined from different sources.

A further important development concerns the work on joint distributions and probabilistic inference in a recursive Bayesian network. Central to this work is the *recursive Markov condition* according to which a variable is conditionally independent of variables that are neither its inferiors nor peers (i.e. occurring at the same level) given its direct superior. The concept of a flattening graph, which is a non-recursive network with arrows from superiors to their inferiors, makes it clear that the recursive Markov condition can be exploited in the same way as the causal Markov condition to yield joint distributions. A related advantage of this condition is that standard inference algorithms for Bayesian networks can be used in the recursive case. This result will be extremely important for further work in this area, particularly in applications. Bayesian networks have received a lot of attention in recent years and the fact that recursive networks can use these techniques will make new developments much more straightforward.

In addition to the substantial developments already mentioned the authors go on to explore the similarity between recursive Bayesian networks and argumentation networks. The idea here is that an argument can be considered as a proposition which supports (or attacks) another proposition. However, an argument can itself support (or attack) another argument and so recursive Bayesian networks can be applied to represent these relationships with arrows now representing the relationship of support (or attack) rather than causality. The recursive nature of these networks can be described more generally in terms of self-fibring networks, where one network is substituted inside another. In this process it is important to ensure

---

[2] A closest common cause C of two nodes A and B is a node for which two causal chains from C to A and C to B have only C in common.

that output from one network is suitable for input to the other. This provides a unified way of looking at recursive causality and similar structures in logic since fibring is a widely applicable approach for combining logics. Thus, two growing areas of artificial intelligence have been brought together in an illuminating way and will undoubtedly lead to further theoretical developments and applications.

In the remainder of this paper I will examine the notion of recursive causality in more detail. In Section 2 I investigate the connection between recursive causality and views concerning the causal relata. I then discuss whether recursive Bayesian networks can be replaced by non-recursive networks in Section 3.

2   Causal Relations as Causes

We return to the fundamental idea of recursive causality and consider whether causal relations can themselves be causes. Can a causal relation such as smoking causes cancer be the cause of some other effect? In attempting to answer this question issues concerning the metaphysics of causality come into play and one's view of the latter will influence their answer to the former. Alternatively, if there are overwhelming reasons for thinking that the answer to the question is 'yes', this may have implications for one's view of causality.

What are the causal relata? What sorts of things can be causes or effects? The standard view is that the causal relata are events, although many alternatives have been proposed. While it is not necessary to go into a detailed discussion concerning the characterisation of events, it is important to note they are generally taken to have a particular spatiotemporal location. It is this concrete nature of events that seems to make them a suitable candidate for the causal relata. For the discussion which follows it is important to distinguish between singular causal claims, such as 'Joe's smoking caused him to develop cancer', from general causal claims, such as 'smoking causes cancer'. According to the event view singular causal claims relate event-tokens, which have a particular spatiotemporal location as discussed above, while general causal claims relate event-types.

The main rival to events as the causal relata are facts. The main objection to facts is that they do not seem to be the right sort of thing to do anything in the world. Unlike events, facts do not have a particular spatiotemporal location and thus seem to be too abstract to push things about in the world. One response to this is to find substitute entities to fulfil this role such as objects [Bennett, 1989], while denying that the causal relation relates these entities. Another response to this argument is that on a Humean concept of causation such entities are not needed and that the regularity can hold between facts. (See [Schaffer, 2000] for further discussion of competing views of the causal relata).

The question of determining whether recursive causality is possible is thus closely related to the question of whether causal relations fall into the same category as the causal relata. In order to explore these issues further three examples of recursive causality will be considered, one of which was discussed by Kim [1976] and two by Williamson and Gabbay.

Case 1.
In his discussion of events as property exemplifications [Kim, 1976, p. 322] asks,

> ... are we to accept these causal relations themselves, i.e. one event's causing another, as events? Or should we fit them into some other ontological category, say, facts?

The action event of Brutus's killing Caesar, for example, seems to be an action of Brutus causing the death of Caesar. If this is correct then the putative event of Brutus killing Caesar is a causal relation between two events: the action of Brutus and the death of Caesar. This causes a problem for Kim since such a complex event does not seem to fit within his framework of events as property exemplifications. Kim's discussion would suggest that causal relations are better thought of as facts than events.

However, as Kim goes on to point out,

> One argument for treating, say, killings as events may be this: they are just the sort of thing that can have causes and effects, and just the sort of thing that can be given causal explanations.

For example, Brutus's killing Caesar may have caused Caesar to be absent from the Roman Senate the next day. This would then be an example of a causal relation being a cause, i.e. an example of recursive causality. Assuming that the causal relata are events this would imply that causal relations are events and so Kim's account of events would be undermined. Kim has a way out, however, since Brutus's killing Caesar involves the event of Brutus's action and the event of Caesar's death and it is surely the latter that causes Caesar's absence the following day. Thus, there is no genuine recursive causality in this case and the argument that actions such as killings (and also causal relations) must be taken to be events is defeated.

Case 2.
As a second possible case of recursive causality, consider the claim that smoking causes cancer which in turn causes the government to restrict tobacco advertising. Note that the effect (i.e. placing a restriction on advertising) is a particular event, an event-token, and so according to the event theory its cause should also be an event-token. Could 'smoking causes cancer' be an event? In particular, could it be an event-token? The claim that smoking causes cancer is a general (rather than singular) causal claim and so it does not have the concrete characteristics of an event-token such as spatiotemporal location.

Perhaps, however, a particular instance of smoking causing cancer could count as an event-token. If so, the general causal claim that smoking causes cancer would then be an event-type. The difficulty with this is that recursive causality would then require that an event-type (smoking causes cancer) could be the cause of an event-token (placing a restriction on advertising). There seem to be two options open: one could deny that the causal relata are events and see if adopting the view that they are facts would help; or one could reject the idea that this is a case of recursive causality. The second option is more straightforward and, as Williamson and Gabbay point out, it is really the *belief* that smoking causes cancer rather than the causal relation itself that causes the ban.

Case 3.
A third possible case of recursive causality is found in the case of preemption: [poisoning causing death] prevents [heart failure causing death]. In order to reject recursive causality Kim's analysis would seem to require that one of the events in the first relation prevents one of the events in the second relation, i.e. either poisoning or death prevents either heart failure or death. None of these possibilities, however, looks promising and so this seems to be a plausible case of recursive causality. If so, one could try to continue to maintain that causal relations are events, although Kim's account of events (and perhaps other accounts) would be undermined. Alternatively, one could claim that the causal relations are facts and so the causal relata must be facts as well.

The idea that causal relations such as heart failure causing death are events is not as problematic in case 3 as it was in case 2. The reason for this is that case 3 would then be a relation between two event-tokens. However, there are other problems with the view that causal relations are events. One advantage of the event view of the causal relata is that there is a concrete entity to do things in the world, but a causal relation seems too abstract for this purpose. Even on a realist view of causality it is difficult to imagine a necessary connection in the world having any role other than in connecting the cause and its effect.

The foregoing discussion suggests that recursive causality does not sit easily with the event view of the causal relata. Thus, if there is genuine recursive causality the causal relations and hence the causal relata are better thought of as facts. Consequently, problems associated with the fact view of the causal relata are inherited. In order to avoid this conclusion, the onus is on the event-theorist to provide a non-recursive account of the proposed examples of recursive causality, especially preemption.

3   Challenges for the Event-theorist

In this section several examples will be investigated from a modelling perspective to determine whether the recursive element can be eliminated from recursive Bayesian networks. This study will not yield a definitive answer to whether there are genuine examples of recursive causality since questions can always be asked about the interpretation of Bayesian networks, i.e. whether they should be interpreted causally and also whether they should be interpreted objectively or subjectively. Nevertheless, if recursive Bayesian networks can be replaced by non-recursive networks this would surely count against recursive causality, whereas if they cannot this would support it.

Causal Relations and Human Actions

Consider again the example of restrictions being placed on advertising as an effect of the causal relation between smoking and cancer. Why does the government decide to make the restrictions? As Williamson and Gabbay point out it is actually the *belief* that smoking causes cancer that gives rise to the action since the restrictions would have been implemented in the absence of the causal relation provided the belief in such a relation was held. In order to model this scenario in a Bayesian network the Belief-Desire-Intention (BDI) model of agency can be used. The BDI model is widely used in artificial intelligence applications and basically involves the idea that an agent adopts certain plans of action (intentions) as a result of beliefs about the world and goals he would like to achieve (desires). A rather simple example of how this could be implemented in a Bayesian network is given in Figure 1, where the belief that smoking causes cancer and the desire to reduce cancer are parents of the node representing the intention to place restrictions on advertising. Of course, in reality, the situation would be more complex since there would be other beliefs, including beliefs about the feasibility of implementing the restrictions, and there may also be conflicting desires.

The network in Figure 1 seems to indicate how recursive causality can be eliminated in this case, but the network does not model how the agent came to hold the belief that smoking causes cancer in the first place. Perhaps it could be argued that a node representing the causal relation between smoking and cancer will be among the parents of the belief node. This claim would not be negated by the fact that the agent could hold a false belief about a causal relation for there may be a probabilistic causal relation such that there is a non-zero probability of believing smoking causes cancer even if it does not. Furthermore, there could be other parents of the belief node which could give rise to a false belief. However, there does not seem to be a convincing reason for thinking that the smoking causes cancer relation must be among the parents. It could be that the parents of the belief node do not include the causal relation, but perhaps statistical information as well as background knowledge. It could then be argued that the causal relation does appear somewhere further up the causal chain rather than as a direct cause of the agent's belief. Thus, it is not obvious that recursive causality can be easily avoided in this case. Nevertheless, it is not a clear cut example of it either and it is certainly the belief in the causal relation that plays the main role.

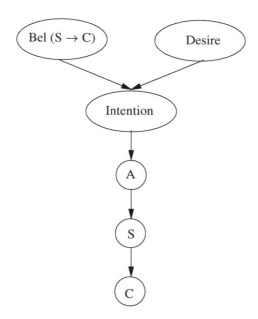

Figure 1. A Belief-Desire-Intention Model in a Bayesian network for the government's decision to restrict tobacco advertising (A) so that it will reduce the level of smoking (S) which will in turn reduce the level of cancer (C).

Prevention
Williamson and Gabbay discuss an example where the taking of mineral supplements prevents [poor diet causing goitre]. They argue that it is not simply the prevention of goitre that is taking place because

> ... there are other causes of goitre such as various defects of the thyroid gland, taking mineral supplements does not inhibit these causal chains and thus does not prevent goitre simpliciter.

Nevertheless, they do point out that in this particular case a non-recursive account can be given since poor diet causes goitre via iodine deficiency, which is prevented by mineral supplements. This is represented in Figure 2 (which is Figure 6 in the Williamson and Gabbay paper).

It also seems, however, that a non-recursive account can be given even in cases where no intermediate variable in the causal chain can be identified. Such an account can be represented by the network given in Figure 3 provided appropriate conditional probabilities are specified. For example, the probability of goitre given poor diet and no supplements would be relatively high, while the probability given

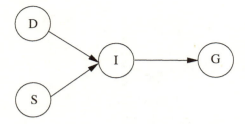

Figure 2. D: poor diet; S: mineral supplements; I: iodine deficiency; G: goitre. Reprinted from the Williamson and Gabbay paper

poor diet and supplements would be relatively low. It could be argued that the causal structure in Figure 2 is much clearer than that in Figure 3 since the latter depends on the probability specifications. However, this is also true in Figure 1 since it is only the probability specification that tells us that supplements prevent (rather than cause) iodine deficiency. A similar approach to that in Figure 3 would enable other causes of goitre not inhibited by supplements to be taken into account. Thus, it seems prevention is not a strong candidate for recursive causality.[3]

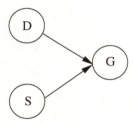

Figure 3. D: poor diet; S: mineral supplements; G: goitre

Preemption
Consider the example, discussed by Lewis [2000], where Suzy and Billy both throw rocks at a bottle. Suzy's rock reaches the bottle first and shatters it, although if Suzy's throw had not taken place Billy's rock would have shattered it instead. This kind of example, referred to as *late preemption*, poses significant problems for many accounts of causality and cannot be represented by a simple Bayesian network with edges from nodes for Suzy's throw (ST) and Billy's throw (BT) to a node for bottle shatters (BS). For example, on a simple counterfactual account Suzy's throw would not be the cause since the effect would have occurred even if she had not thrown the rock. One approach to avoid this problem is to intro-

---

[3] A similar analysis can be given in the case of context specificity.

duce intermediate variables. Halpern and Pearl [2001], for example, represent this scenario by the network in Figure 4 with intermediate nodes BH, Billy's rock hits the (intact) bottle, and SH, Suzy's rock hits the (intact) bottle. Note that the link between SH and BH is one of prevention. By adopting this approach they are able to establish that Suzy's throw is the actual cause of the bottle shattering (see also [Pearl, 2000]).

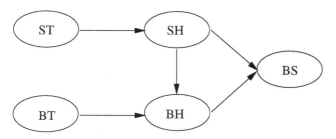

Figure 4. ST: Suzy throws a rock at the bottle; BT: Billy throws a rock at the bottle; SH: Suzy' s rock hits the bottle; BH: Billy's rock hits the bottle; BS: bottle shatters

A problem with such accounts, however, arises in cases of *trumping* [Schaffer, 2000] in which there are no plausible intermediaries. By accepting recursive causality a more straightforward approach is available which does not resort to intermediate variables. A recursive network for the example discussed above is given in Figure 5 where it should be noted that the link between the nodes is one of prevention. This example appears to be a case where eliminating the recursive element would only give rise to problems and thus provides a good reason for accepting the notion of recursive causality.

Figure 5. ST→BS: Suzy's throw causes the bottle to shatter; BT→BS: Billy's throw causes the bottle to shatter

4 Conclusions

The discussion in Sections 2 and 3 has looked at the question of whether causal relations can themselves be causes. In some cases which might seem to involve recursive causality the recursive element can be eliminated, but the case of preemption seems to be a strong candidate for a genuine case of recursive causality. It is well-known that preemption raises problems for various accounts of causality so the fact that recursive causality handles it so easily counts in its favour. However, I have also suggested, following Kim, that there are links between one's view of the causal relata and recursive causality. The suggestion is that if causal relations can be causes, then the causal relata are better thought of as facts. Thus, there is a dilemma for the defender of the event view of the causal relata: either all examples of recursive causality should be rejected which is undermined in the case of preemption, or the event view should be rejected.

Acknowledgements

I would like to thank Dr Jon Williamson and Prof Dov Gabbay for helpful discussions.

BIBLIOGRAPHY

[Bennett, 1989] J. Bennett. *Events and their Names*. Oxford: Clarendon Press, 1989.
[Halpern and Pearl, 2001] J. Halpern and J. Pearl. Causes and Explanations: a Structural-Model Approach - Part I: Causes. *Proceedings of the 17th Conference on Uncertainty in AI,* pp. 194–202, 2001.
[Kim, 1976] J. Kim. Events as Property Exemplifications. In M. Brand and D. Walton, eds. , *Action Theory*, pp 159–177. Dordrecht: D. Reidel Publishing, 1976. Reprinted in S. Laurence and C. Macdonald, eds., *Contemporary Readings in the Foundations of Metaphysics*, pp. 310–326. Oxford: Blackwell, 1988.
[Lewis, 2000] D. Lewis. Causation as Influence. *Journal of Philosophy*, **97**, 182–97, 2000.
[Pearl, 2000] J. Pearl. *Causality: Models, Reasoning and Inference*. Cambridge: Cambridge University Press, 2000.
[Schaffer, 2000] J. Schaffer. Trumping Preemption. *Journal of Philosophy*, **97**, 165–81, 2000.
[Schaffer, 2003] J. Schaffer. The Metaphysics of Causation, *The Stanford Encyclopedia of Philosophy* (Spring 2003 Edition), Edward N. Zalta (ed.), URL = http://plato.stanford.edu/archives/spr2003/entries/causation-metaphysics/

David H. Glass

School of Computing and Mathematics
University of Ulster, Newtownabbey
Co. Antrim, BT37 0QB, UK
Email: dh.glass@ulster.ac.uk

On Gabbay's Fibring Methodology for Bayesian and Neural Networks
A Comment on Williamson and Gabbay

A. S. D'AVILA GARCEZ

1 Introduction

Fibring is a methodology for combining logics by breaking, manipulating and re-arranging them into simple components. Briefly, the languages and inference rules of each logic may appear in the combined system, while the semantics of the fibring needs to be carefully compiled as a combination of the classes of models of the logics. For example, in the case of the fibring of a propositional temporal logic with a propositional linear space logic, each model of the fibring will contain a time line and a space line [Sernadas *et al.*, 1999].

In addition to its appeal from a purely logical point of view, fibring has a number of applications in mathematics and computer science. It has important implications to knowledge representation and artificial intelligence—where, for example, the combination of temporal and deontic logics may be necessary—and to formal methods and software engineering, where one may need to work with declarative specifications and procedural ones. Putting together artificial intelligence and software engineering, in robotics, for instance, a robot's visual system may require one logical representation, while its planning system may require another [Gabbay, 1999]. Moving even further, some of the robot's systems may not be a logical (symbolic) system, but a connectionist (sub-symbolic) one, as in the case of neural networks systems very successfully used for visual information processing, or Bayesian networks also successfully used to model uncertainty. The concept of fibring, therefore, needs to be extended to cater for hybrid systems. It may, how-ever, continue to have logic as its underlying mechanism with the help of neural-symbolic learning systems [d'Avila Garcez *et al.*, 2002], as we shall exemplify in the sequel.

In what follows, we discuss how Bayesian networks may be self-fibred. We then discuss how Neural Networks are fibred, and exemplify the fibring of Bayesian and neural networks using an argumentation framework example in law. We conclude by discussing how neural-symbolic learning systems may serve as the underlying framework for the fibring of logics and networks.

Laws and Models in in Science, 233–245.

2   Fibring Bayesian Networks

The idea of fibring Bayesian networks emerged from the observation that causal relations may themselves take part in causal relations. This has been simply and effectively exemplified in the very first section of Williamson and Gabbay's *Recursive Causality in Bayesian Networks and Self-Fibring Networks* [Williamson and Gabbay, 2004]. The example states that the fact that smoking causes cancer, for instance, causes the government to restrict tobacco advertising. Such a recursive definition of causal relations was then used to define what Williamson and Gabbay call a *recursive Bayesian network*, a Bayesian network in which certain nodes may be Bayesian networks in their own right. This is defined with the help of the concept of *network variables,* which are variables that may take Bayesian networks as values. Thus, a network variable SC may be used to represent the fact that *smoking causes cancer*, and then SC causes A, where A stands for *restricting advertising*. This may be written as $SC \to A$ where $SC = S \to C$, or simply as $(S \to C) \to A$.

A recursive network can be interpreted as a non-recursive network if one treats network variables SC as simple variables SC'. In addition, a recursive network can be *flattened* into its non-recursive counterpart, given a consistent assignment of values v to the domain V of the recursive network. Consistency of assignments is, therefore, a requirement for the use of standard Bayesian networks algorithms by recursive Bayesian networks. This is why Williamson and Gabbay devote an entire section of their paper to the issue of consistency. They identify three forms of consistency between networks: causal consistency (w.r.t. causal relations), Markov consistency (w.r.t. implied probabilistic independencies), and probabilistic consistency (w.r.t. probability functions).

Note that variables are allowed to occur more than once in a recursive network, e.g., $(A \to B) \to A$. If we are not careful, its flattened counterpart network may have cycles, what would prevent the use of standard Bayesian inference algorithms. Williamson and Gabbay concentrate, as a result, on consistent acyclic assignments. It is worth noting also that there exists no straightforward flattening of a recursive Bayesian network into a Bayesian network when no values v are given, i.e. flatenings are relative to an assignment v on the domain V.

Once the flattening of recursive Bayesian networks is fully defined (allowing for the use of a standard Bayesian network to determine the probability distributions), Williamson and Gabbay investigate the more general concept of allowing graphs inside graphs, referred to as *self-fibring of networks*. They present a generalisation of recursive Bayesian networks in the form of an important definition of fibred networks in terms of a fibring function, which is general enough to comprise the fibring of neural networks as well [Haykin, 1999], as discussed in the following section. This definition is obtained by looking at the arrows in directed graphs as different types of implications in logic such as substructural or intuitionistic implication, Dempster–Shafer or causal Bayesian implication. In this general setting, the fibring function is responsible for providing the different interpretations; in the case of Bayesian networks, the fibring function depends on a table of conditional probabilities.

The general case problem of self-fibring of networks can be defined as follows: Let $\mathbf{B}(X)$ be a network containing a node X. Let $\mathbf{A}$ be a network. Find $\mathbf{C}=\mathbf{B}(X/A)$, a new network which is the result of substituting $\mathbf{A}$ for X in $\mathbf{B}$. The power of fibring, however, lies in the fact that this substitution is not only performed on the syntactical level, but takes the form of a *semantic insertion*, which makes use of the fibring function as follows. The substitution of $\mathbf{A}$ takes into account the meaning of $\mathbf{B}$. For example, if $\mathbf{B}$ is a Bayesian network in which X may take two values, say $\{0,1\}$, then if X is 0 we substitute $\mathbf{A}_0$, and if X is 1 we substitute $\mathbf{A}_1$. We write $\mathbf{C}= f_x(\mathbf{A},\mathbf{B})$, where f is our fibring function.

It is the concept of semantic insertion that makes fibring a powerful mechanism. For example, consider the network of Figure 1. The output of fibring function f is allowed to modify the fibring function g itself. In the case of Bayesian networks, where f and g are probability matrices, one could decide to modify g by, say, multiplying it by f. The resulting network would be: $a \overrightarrow{f} (b \overrightarrow{f.g} c)$.

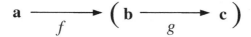

Figure 1. A recursive network $a \to X$ where $X = b \to c$.

In the case of neural networks, where f and g are weight vectors, the idea of multiplying f by g is quite natural since this is already what non-recursive neural networks do. We shall see in the sequel that applying the concept of semantic insertion to neural networks results in recursive networks being strictly more expressive than non-recursive ones, in the sense that they cannot be flattened into equivalent non-recursive networks, even when the straightforward *multiplication rule* is used for fibring.

3   Fibring Neural Networks

The goal of Neural-Symbolic integration [Cloete and Zurada, 2000; d'Avila Garcez *et al.*, 2002] is to benefit from the combination of the symbolic and the connectionist paradigms of Artificial Intelligence. To this end, we know that a fundamental aspect of symbolic computation lies in the ability to perform recursion. Recursion in neural networks, then, is the idea of allowing networks to be composed not only of interconnected neurons but also of other networks (called embedded networks). But the idea is not simply to organise networks as a number of sub-networks. Borrowing the concept of semantic insertion from fibring, the function computed by an embedded network may depend on the function computed by the embedding network.

A neural network consists of interconnected neurons (or processing units) that compute a simple function according to the weights (real numbers) associated to the connections. Learning in this setting is the incremental adaptation of the weights. The interesting characteristics of neural networks do not arise from the simple functionality of each neuron, but from their collective behaviour. In the case of recursive neural networks, introduced in [d'Avila Garcez and Gabbay, 2004], we see fibring as learning, where the change of the weights of a neural network implies the change of the function computed by it. So, we simply use the weights of the embedding network to change the weights of the embedded network. When running the network $a \overrightarrow{f} (b \overrightarrow{g} c)$, we obtain the network $a \overrightarrow{f} (b \overrightarrow{f.g} c)$, where a, b and c are neurons, and f and g are weights. In other words, *one network is being used to train the other*, as the following example illustrates.

EXAMPLE 1. Consider the networks **A** and **B** of Figure 2, each with inputs i_1 and i_2, and output o_1. Let network **B** be embedded into the output neuron of network **A**, as shown in the figure. This indicates that the state of **A**'s output neuron (**s**) will influence the weights of **B** $(W_1^\mathbf{B}, W_2^\mathbf{B}, W_3^\mathbf{B})$, according to a fibring function (φ). Using the simple multiplication rule for φ, let us take $\mathbf{W}_{new}^\mathbf{B} = \mathbf{s} \cdot \mathbf{W}_{old}^\mathbf{B}$. Notice how network **B** is being trained (when φ changes its weights) at the same time that network **A** is running.

Fibred neural networks can be trained from examples in the same way that standard feedforward networks are (for example, with the use of Backpropagation [Rumelhart *et al.*, 1986]). Networks **A** and **B** of Figure 2, for example, could have been trained separately before being fibred. Network **A** could have been trained, e.g., with a robot's visual system, while network **B** would have been trained with its planning system. For now, we assume that, once defined, the fibring function itself remains unchanged. Future extensions of fibring neural networks could, however, consider the task of learning fibring functions as well.

In addition to using different fibring functions, networks can be fibred in a number of different ways as far as their architectures are concerned. The networks of Figure 2, for example, could have been fibred by embedding Network **B** into an input neuron of Network **A** (say, the one with input i_1), thus changing the value of **s** (to the state of this input neuron) used to calculate the new set of weights of **B**, and thus also changing the function computed by **B**.

The recursive network of Figure 2 is capable of computing, with the use of the multiplication rule for fibring, an output n^2 for $n \in \Re$ given as input. In other words, recursive networks are capable of performing the exact computation of the square of their input for any input in $\Re$. This indicates that recursive networks approximate functions in an unbounded domain [Henderson, 2002;

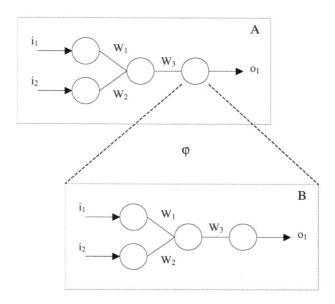

Figure 2. Fibring two simple networks

Hines, 1996], as opposed to non-recursive feedforward networks. In fact, in addition to being universal approximators, recursive networks can approximate any polynomial function, and thus are more expressive than non-recursive feedforward neural networks. The proofs are given in [d'Avila Garcez and Gabbay, 2004].

Finally, note that many networks can be embedded into a single network, and that networks can be nested so that network B is embedded into network A, network C is embedded into network B, and so on. In addition to the existing multitude of fibring functions, there is a multitude of network architectures for fibring that might be interesting investigating. The choice of fibring function and architecture is domain dependent, and an empirical evaluation of recursive networks in comparison with standard neural networks would also be required. Another interesting open question is that of which logics could be represented by recursive networks. The extra expressiveness of such networks contributes to the development of the research on Neural-Symbolic integration, where neural networks need to be used to perform complex symbolic computation.

REMARK 2. In summary, both recursive Bayesian networks and recursive neural networks share the more general principles and definitions of Gabbay's fibring methodology. However, each technique has very specific characteristics and algorithms to deal with uncertainty (in the case of Bayesian networks) and learning and generalisation capabilities (in the case of neural networks). Both recursive Bayesian and neural networks are strictly more expressive than their standard versions, not only because of the idea of allowing networks inside networks, but also because of the general concept of semantic insertion from fibring. As we have seen, recursive Bayesian networks when flattened may produce networks with cycles instead of standard Bayesian networks, while recursive neural networks are capable of computing unbounded functions exactly, as opposed to standard networks. Standard Bayesian networks algorithms can be applied directly to the component networks of a recursive Bayesian network, and under certain conditions to the recursive network itself. Similarly, standard neural networks learning algorithms may be applied directly to each of the component networks of a recursive neural network. Finally, both types of networks may have counterpart symbolic representations with the help of Labelled Deductive Systems [Gabbay, 1996] and Neural-Symbolic Learning Systems [d'Avila Garcez et al., 2002], and these logical representations may be helpful in guiding the development of the research on fibring networks.

4   Fibring Applied to Argumentation

An interesting application of fibring is in the area of legal reasoning and argumentation [d'Avila Garcez et al., 2004; Gabbay and Woods, 2003]. In [Gabbay and Woods, 2003], for example, Gabbay and Woods argue for the combined use of Labelled Deductive Systems and Bayesian networks to support legal evidence reasoning under uncertainty. In addition, they argue that neural networks, as learning systems, could play a role in this process by being used to update/revise degrees of belief and the rules of the system whenever a new evidence is presented. The three different representations all expand a value-based argumentation framework outlined in [Bench-Capon, 2003], in which argumentation networks are used to model arguments and counter-arguments. A typical example in the area is the following *moral debate* example.

> Hal, a diabetic, loses his insulin in an accident through no fault of his own. Before collapsing into a coma, he rushes to the house of Carla, another diabetic. She is not at home, but Hal breaks into her house and uses some of her insulin. Was Hal justified? Does Carla have a right to compensation?

The following are some of the arguments involved as presented in [Bench-Capon, 2003].

A: Hal is justified, he was trying to save his life;
B: It is wrong to infringe the property rights of another;
C: Hal compensates Carla;

D: Hal is endangering Carla's life; and
E: Carla has abundant insulin.

In [Bench-Capon, 2003], arguments and counter-arguments are arranged in an argumentation network, as in Figure 3, where an arrow from argument X to argument Y indicates that X attacks Y. For example, the fact that it is wrong to infringe Carla's right of property (**B**) attacks Hal's justification (**A**).

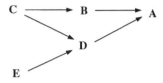

Figure 3. Part of the argumentation network for the moral debate example

Some aspects of the argumentation network of Figure 3 are probabilistic. For example, the question of whether Carla has abundant insulin (**E**) depends on the time and is a matter of probability. The question of whether Hal will be able to compensate Carla with replacement insulin in time (**C**) is also a matter of probability. If Carla has abundant insulin, the chances that Hal will be able to compensate her are higher. The probability matrices of this Bayesian network (**E**→**C**) influence whether Hal is endangering Carla's life by stealing some of her insulin (**D**). In the same argumentation network, some other aspects may change as the debate progresses and actions are taken; the strength of one argument in attacking another may change in time. This is a learning process that can be implemented using a neural network in which the weights record the strength of the arguments. The neural network for the set of arguments {**A**, **B**, **D**} is depicted in Figure 4.

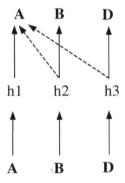

Figure 4. A neural network for arguments {A,B,D}

The neural network of Figure 4 is an auto-associative single hidden layer network with input (A,B,D), output (A,B,D) and hidden layer (h1,h2,h3). Solid arrows represent positive weights and dotted arrows represent negative weights. Arguments are supported by positive weights and attacked by negative ones. Argument **A** (input neuron **A**), for example, supports itself (output neuron **A**) with the use of hidden neuron h1. Similarly, **B** supports itself (via h2), and so does **C** (via h3). From the argumentation network, **B** attacks **A**, and **D** attacks **A**. The attacks are implemented in the neural network by the negative weights (see dotted lines in Figure 4) with the use of h2 and h3.[1] The network of Figure 4 is a standard feedforward neural network that can be trained, e.g., with the use of the standard Backpropagation learning algorithm [Rumelhart *et al.*, 1986]. Training would change the initial weights of the network (the initial belief on the strength of arguments and counter-arguments, which could be random), according to examples of input/output patterns, i.e. examples of the relationship between arguments **A**, **B** and **D**. Roughly speaking, if then the absolute value of the weight from neuron h1 to output neuron A is greater than the sum of the absolute values of the weights from neurons h2 and h3 to **A**, one can say that argument **A** prevails (in which case output neuron **A** should be *activated* in the neural network).

The key to running the network properly is to connect output neurons to their corresponding input neurons using weights fixed at 1, so that the activation of output neuron **A**, for example, is fed into the activation of input neuron **A** the next time round. This implements chains such as A attacks B, B attacks C, C attacks D, and so on, by propagating activations around the network. The following example illustrates the dynamics of argumentation neural networks (see [d'Avila Garcez *et al.*, 2004] for details).

EXAMPLE 3. Take the case in which an argument A attacks an argument B, and B attacks an argument C, which in turn attacks A in a cycle. In order to implement this in a neural network, we need three hidden neurons (h1, h2, h3), positive weights to explicitly represent the fact that A supports itself (via h1), B supports itself (via h2), and so does C (via h3). In addition, we need negative weights from h1 to B, from h2 to C and from h3 to A to implement attacks. If all the weights are the same in absolute terms, no argument wins, as one would expect, and the network stabilises with none of output neurons {A,B,C} activated. If, however, the value of argument A (i.e. the weight from h1 to A) is stronger than the value of argument C (the weight from h3 to C, which is expected to be the same in absolute terms as the weight from h3 to A), C cannot attack and defeat A. As a result, A is activated. Since A and B have the same value, B is not activated, since the weights from h1 and h2 to B will both have the same absolute value. Finally, if B is not activated then C will be activated, and a stable state {A,C} will

---

[1] The hidden neurons are used in the network to provide a greater flexibility as to what can be learned as combinations of the input neurons.

be reached in the network. In Bench-Capon's model [Bench-Capon, 2003], this is precisely the case in which colour blue is assigned to A and B, and colour red is assigned to C with blue being stronger than red. Note that the order in which we reason does not affect the final result (the stable state reached). For example, if we started from B successfully attacking C, C would not be able to attack A, but then A would successfully attack B, which would this time round not be able to successfully attack C, which in turn would be activated in the final stable state {A,C}. This indicates that a neural implementation of this reasoning process may, in fact, be advantageous from a purely computational point of view due to neural networks' parallel nature.

Now that we have two more concrete models of the arguments involved in the moral debate example—a probabilistic model and a learning/action model—we can reason about the problem at hand in a more realistic way. We just need to put the two models together with the use of the fibring methodology for networks. The (more abstract) argumentation network of Figure 3 can be used to tell us how the networks (Bayesian and neural) are to be fibred. From Figure 3, one can see that both arguments C and E attack argument D directly. As a result, we would like the probabilities in our Bayesian network $E \to C$ to influence the activation of neuron D in the neural network. Thus, network $E \to C$ needs to be embedded into node D. Again from the argumentation network (Figure 3), one can see that argument C also attacks argument B directly. As a result, we would like the probabilities associated with C to influence the activation of neuron B. As before, this can be done by embedding Bayesian network C into neuron B. This produces the recursive network of Figure 5.

Let us consider again the embedding of $E \to C$ into D. We have seen that the embedding is guided by the arrow in the original argumentation network. The arrow in an argumentation network indicates an *attack*. As a result, the higher the probability P(C/E) in $P2$ is (see Figure 5), the lower the activation value of neuron D should be. Similarly, the higher the probability P(C) in $P1$, the lower the value of B should be. Thus, we take $\varphi 1 : s^{\mathbf{B}}_{new} = s^{\mathbf{B}}_{old} - $ P($\mathbf{C}$) and $\varphi 2 : s^{\mathbf{D}}_{new} = s^{\mathbf{D}}_{old} - $ P($\mathbf{C/E}$), where P($\mathbf{X}$) $\in [0,1]$ and $\mathbf{s} \in (0,1)$. This definition of the fibring functions completes the fibring.

In this case study, we have seen that the fibring of networks can be used to combine different systems, yet maintaining their individual characteristics and using their algorithms for learning and reasoning. In the combined system, the new state of output neuron D ($s^{\mathbf{D}}_{new}$) will then be fed into input neuron D and affect the new state of A (through hidden neuron h3 and the negative weight from h3 to A), such that the higher the value of D the lower the value of A. The same will happen through B according to the dynamics of the embedding and embedded networks, and this will allow the reasoning as to whether Hal is justified or not to proceed.

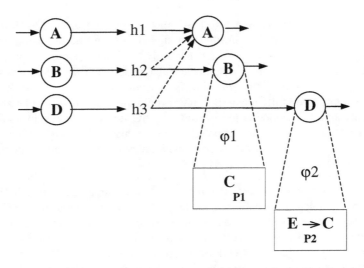

Figure 5. Fibring of Bayesian and neural networks applied to value-based argumentation

5   Conclusion

We have seen that the general methodology of fibring logical systems can be applied also to sub-symbolic systems such as Bayesian and neural networks. As in symbolic systems, where for example the fibring of two logics without embedded implication may result in a logic with embedded implication, the fibring of networks may result in embedded networks such as $(A \to B) \to C$, which are strictly more expressive than standard networks (i.e.: do not have a flattened counterpart network). This indicates, now from the point of view of neural-symbolic integration, that fibring may be used to produce simple neural network architectures (an important requirement for effective neural networks learning) that represent powerful logics such as modal, temporal, first order and higher order logics.

As an example, consider the network structure of Figure 6. The small networks inside each agent may contain an intricate architecture used to represent the knowledge of the agent at a time point. These networks may relate to each other in a metalevel network where, in the horizontal axis, the knowledge of the different agents at a time point is presented, and in the vertical axis, the evolution of the agents' knowledge through time is presented. This yields a distributed, massively parallel, multi-agent system, created out of recursive networks, in which space and time logics are incorporated (see [d'Avila Garcez and Lamb, 2004] for details). This illustrates how the general fibring methodology can be applied to the integration of symbolic and sub-symbolic systems.

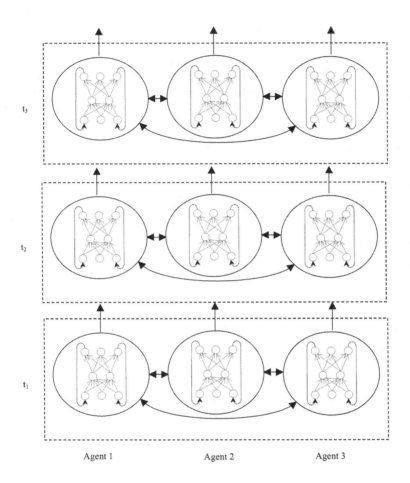

Figure 6. Fibring of knowledge and time in neural networks

Neural-Symbolic Learning Systems [d'Avila Garcez *et al.*, 2002], therefore, by providing translation algorithms and proofs of the correctness of such translations between different neural networks and different logics, may serve as the underlying framework for the progress of the study of fibring between networks and logics, and of the self-fibring of networks. In this setting, an appropriate fibring of two networks A and B, for example, would be one in which the logic extracted from the fibred network is the same as the logic obtained from fibring the logics extracted from networks A and B, respectively. There are many avenues of research on the self-fibring of networks and, more generally, on the integration of logics and networks. In this paper, I hope to have convinced the reader that a way forward to develop the research on Artificial Intelligence is to study hybrid systems in conjunction with Gabbay's fibring methodology.

BIBLIOGRAPHY

[Bench-Capon, 2003] T. J. M. Bench-Capon. Persuasion in practical argument using value-based argumentation frameworks. *Journal of Logic and Computation*, 13:429–448, 2003.

[Cloete and Zurada, 2000] I. Cloete and J. M. Zurada, editors. *Knowledge-Based Neurocomputing*. The MIT Press, 2000.

[d'Avila Garcez and Gabbay, 2004] A. S. d'Avila Garcez and D. Gabbay. Fibring neural networks. In *Proc AAAI'04*, San Jose, USA. AAI Press, 2004.

[d'Avila Garcez and Lamb, 2004] A. S. d'Avila Garcez and L. C. Lamb. Reasoning about time and knowledge in neural-symbolic learning systems. In L. Saul S. Thrun and B. Schoelkopf, editors, *Advances in Neural Information Processing Systems 16*, Proceedings of the NIPS 2003 Conference, Vancouver, Canada, MIT Press, 2004.

[d'Avila Garcez et al., 2002] A. S. d'Avila Garcez, K. Broda, and D. M. Gabbay. *Neural-Symbolic Learning Systems: Foundations and Applications*. Perspectives in Neural Computing. Springer-Verlag, 2002.

[d'Avila Garcez et al., 2004] A. S. d'Avila Garcez, D. M. Gabbay, and L. C. Lamb. Value-based argumentation as neural networks. *forthcoming*, 2004.

[Gabbay and Woods, 2003] D. M. Gabbay and J. Woods. The law of evidence and labelled deduction: A position paper. *Phi News*, 4, October 2003.

[Gabbay, 1996] D. M. Gabbay. *Labelled Deductive Systems*, volume 1. Clarendom Press, Oxford, 1996.

[Gabbay, 1999] D. M. Gabbay. *Fibring Logics*. Oxford Univesity Press, 1999.

[Haykin, 1999] S. Haykin. *Neural Networks: A Comprehensive Foundation*. Prentice Hall, 1999.

[Henderson, 2002] J. Henderson. Estimating probabilities of unbounded categorization problems. In *Proceedings of European Symposium on Artificial Neural Networks*, pages 383–388, Bruges, Belgium, April 2002.

[Hines, 1996] J. Wesley Hines. A logarithmic neural network architecture for unbounded non-linear function approximation. In *Proceedings of IEEE International Conference on Neural Networks*, Washington, USA, June 1996.

[Rumelhart et al., 1986] D. E. Rumelhart, G. E. Hinton, and R. J. Williams. Learning internal representations by error propagation. In D. E. Rumelhart and J. L. McClelland, editors, *Parallel Distributed Processing: Explorations in the Microstructure of Cognition*, volume 1, pages 318–362. MIT Press, 1986.

[Sernadas et al., 1999] A. Sernadas, C. Sernadas, and C. Caleiro. Fibring of logics as a categorical construction. *Journal of Logic and Computation*, 9(2):149–179, 1999.

[Williamson and Gabbay, 2004] J. Williamson and D. Gabbay. *Recursive Causality in Bayesian Networks and Self-Fibring Networks*. This volume. 2004.

Artur D'Avila Garcez

Department of Computing
School of Informatics
City University London
London EC1V 0HB, UK
Email: aag@soi.city.ac.uk

Philosophy of Science in a European Perspective

MARIA CARLA GALAVOTTI

An Overview of the activities of the European Science Foundation Network 'Historical and contemporary perspectives of philosophy of science in Europe'

The activities of the ESF Network on 'Historical and contemporary perspectives of philosophy of science in Europe' began in October 2000, with a meeting of its Coordinating Committee, held in Strasbourg on the premises of the European Science Foundation. That first meeting, having the purpose of planning the three-years life period of the Network, was very fruitful because after an intensive exchange of email messages over several months, we finally had the chance of sitting at the same table for two days, to discuss our objectives and plan our activities in some detail. The discussion was very constructive, not only did it point to a common ground of ideas and goals, but it created a relaxed and cooperative atmosphere, that accompanied the Coordinating Committee ever since. During the first meeting, a text was compiled, describing the aims of the Network, to appear on a flyer, as well as in the Network web-site. It describes the main goal of the Network as that of 'bringing together scholars in different European Countries, in the hope of identifying the main features and topics of philosophy of science in Europe and re-establishing the flourishing European intellectual and philosophical activity that existed early in the 20th century'.

To achieve that goal, the Network organized three workshops, whose proceedings will eventually appear in three books. The subjects of these workshops, chosen in view of their significance to the current debate in philosophy of science, were: 'Observation and experiment in the natural and social sciences', 'Induction and deduction in the sciences', and 'Laws and models in science'. We decided that each workshop would include between 8 and 10 invited talks, followed by one or two invited comments—in fact we started with the idea of having two commentators, but after the first workshop we opted for having just one, to leave more time for discussion.

Laws and Models in Science, 247–252.

The first workshop on *Observation and experiment in the natural and social sciences* took place at the Conference Centre of Bertinoro, near Bologna, from the 30th of September to October 2nd, 2001. It hosted 8 main papers, followed by comments. The speakers were Patrick Suppes, with comments by Laszlo Szabo and Paolo Legrenzi; Giora Hon, with comments by Gereon Wolters; Ursula Klein, with comments by Donald Gillies and Aristides Baltas; Reinhard Selten, with comments by Wenceslao Gonzalez and Roberto Scazzieri; David Atkinson, with comments by Michael Stöltzner and Daniel Andler; Gerd Gigerenzer, with comments by Jeanne Peijnenburg and David Papineau; Paul Schoenle, with comments by Roberto Festa; Colin Howson, with comments by Ilkka Niiniluoto and Paolo Garbolino. The proceedings of this conference are published in a volume bearing the same title of the conference, edited by Maria Carla Galavotti for the 'Boston Studies in the Philosophy of Science' series (Kluwer, 2003). The book includes all the papers presented at the conference, except that of Paul Schoenle (and Roberto Festa's comments). Additional comments by Igor Douven (commenting on Howson), Gurol Irzik (commenting on Klein), Miklos Redei (commenting on Atkinson) and Raffaella Campaner (commenting on Hon) were included.

The second workshop on *Induction and deduction in the sciences* was held in Vienna on July 7th–9th, 2002. It hosted 11 papers, each followed by one commentary. The speakers were Theo Kuipers, with comments by Adam Grobler; Joke Meheus, with comments by Matti Sintonen; Hans Rott, with comments by Nils-Eric Sahlin; Wolfgang Spohn, with comments by Eckehart Köhler; Friedrich Stadler, with comments by Malachi Hacohen; Peter Clark, with comments by Jean-Jacques Szczeciniarz; Karl Milford, with comments by David Miller; Ivor Grattan-Guinness, with comments by Ladislav Kvasz; Ilkka Niiniluoto, with comments by Stathis Psillos; Donald Gillies, with comments by Stephan Hartmann; and Ryszard Wojcicki, with comments by Max Urchs. The proceedings of this conference appear in a volume bearing the title of the workshop, edited by Friedrich Stadler, as part of the series *Vienna Institute Yearbook* (Kluwer 2004).

The third conference on *Laws and models in science* was held in London, on the premises of King's College, on September 7th–9th, 2003, and its proceedings are published in the present volume.

When organizing the Network workshops, a main concern of the Coordinating Committee has been to bring together philosophers and scientists, especially from areas like economics, psychology, brain research, artificial intelligence, which are more 'new' to philosophy of science than other disciplines, like physics, traditionally involved with foundational problems. People like Reinhard Selten (a Nobel laureate in economics), Paul Schoenle (a neurologist), Gerd Gigerenzer (a cognitive scientist) and Ursula Klein (a historian of chemistry) were invited to Bertinoro precisely in this spirit, and so were the experimental psychologist Paolo Legrenzi and the economist Roberto Scazzieri. At the same time, physics was not neglected, as testified by the papers of Atkinson and Suppes, and the comments of Szabo. The Vienna workshop did not accommodate an analogous proportion of philosophers and scientists because its topic called for a stronger presence of logicians, or logically-oriented philosophers of science, rather than scientists on the field. Therefore, the workshop hosted a number of papers centred on formal methods and their use in the philosophy of science, including those of Kuipers, Niiniluoto, Meheus, Rott and Spohn. In addition, various historical contributions, like those of Stadler, Clark, Milford and Grattan-Guinness, were presented at the Vienna gathering. At the last conference in London, we wanted to have again a number of contributions dealing with problems related more directly to science, and in this spirit we invited Shallice, Hegselmann, Gabbay, and Butterfield.

The conviction that philosophy of science should be backed by a constant dialogue between philosophers and scientists imprinted the beginnings of philosophy of science in the last century—as testified by the 'Manifesto' of the Vienna Circle, the programmes of the conferences organized by logical empiricists, or the first volumes of *Erkenntnis*. Today this practice is no longer so widespread, except in some areas, like logic and philosophy of physics, which have become, or are on the verge of becoming autonomous fields—in fact logic has long ago split into mathematical and philosophical logic, and the existence of a separate ESF Network on philosophy of physics is a clear sign that philosophy of physics is moving in the same direction. The members of the Coordinating Committee of the Network agreed in regarding the dialogue between philosophers and scientists as indispensable to a philosophy of science done 'from within', and this confers an extra-value to the Network workshops, and to the volumes of their proceedings. Hopefully, the work done by the Network has and will incline scientists to look more confidently at philosophy of science, and fostered a fruitful exchange between scientists and philosophers. This kind of exchange is something philosophers of science should strive for, because it is only through a constructive dialogue with scientists that philosophy of science can be done in an insightful way.

A further conviction shared by the members of the Coordinating Committee was that a responsible and successful approach to philosophy of science must not forget about the history of the concepts it employs. This has been another leading idea in the organization of the three workshops, all of which hosted historical papers. Indeed, also presentations that cannot be deemed historical in a strict sense, showed a peculiar historical awareness. This is another distinctive feature of the final products of the Network.

A major concern of the Coordinating Committee when organizing the workshops has been that of bringing together people from the widest range of Countries, especially in the area covered by the ESF, without of course sacrificing quality. In addition, a few guests from non-European Countries were invited. Just to give some data: researchers from the following Countries convened at the three workshops: Australia (1), Austria (4), Belgium (3), Finland (2), France (4), Germany (12), Great Britain (15), Greece (2), Holland (6), Hungary (2), Israel (1), Italy (6), Poland (4), Slovakia (1), Spain (2), Sweden (2), Turkey (1), USA (2)—it makes for 18 Countries, which become 20 if we consider England, Scotland and Northern Ireland separately... as done by the European Football Association. One can say that exchange among scholars from all over Europe has been encouraged to the point that most, if not all, those European Countries in which philosophy of science is a topic of some concern have been represented at the workshops.

No doubt, the quality of the workshops has been first rate. Both the papers that were read and the extensive discussion they provoked are evidence that philosophy of science in Europe is not only very lively, but teems with original and seminal ideas. Did the Network succeed in pointing out a European way to philosophy of science? An answer to this question involves philosophical remarks, that will inevitably mingle with some personal considerations. The following observations, though, are meant as a matter for discussion and future work.

One might find it odd that the first conference of a ESF Network was opened by an American, namely Patrick Suppes. This choice was meant as a homage to one of the 'established greats' of philosophy science, and to the pluralism that imprinted his work. When (more than 40 years ago) Suppes started calling the attention of philosophers of science to the role of models in science—especially of 'models of data'—and urged the need to open philosophy of science to the experimental aspects of scientific research, he enacted within philosophy of science a revolution no less important than the more celebrated 'Kuhnian revolution'. To be sure, the preceding remark is not meant to detract from the work of Thomas Kuhn, who exercised a tremendous impact on philosophy of science. Rather, it is meant to suggest that the import of Suppes' move is still underestimated, when in fact his work impressed on philosophy of science a decisive turn.

Opening philosophy of science to the 'context of discovery'—to use Reichenbach's well known terminology—yields important consequences. In the first place, it makes observation and experimentation objects of enquiry of their own. Furthermore, it enforces a reconsideration of a number of crucial issues, including the structure of scientific theories, the relationships between theoretical and observational language, the nature and role of formal methods. The analysis of such issues needs a more powerful conceptual machinery than that employed by logical empiricists. In particular, the syntactical and semantical methods of (traditional) symbolic logic have to be supplemented by probabilistic inferential methods, and by the statistical methodology adopted at the various stages of scientific investigation, from the collection and analysis of data to the assessment and testing of hypotheses. This change of perspective brings models (ranging from experimental to theoretical models) to the core of philosophy of science, and gives way to a novel approach towards methodological and foundational issues, characterized by a peculiar pluralism.

The above mentioned conceptual innovations were discussed at great length at the workshops organized by the Network, where a genuinely pluralistic attitude inspired both the papers that were read, and the discussion they provoked. Pluralism looks like a distinctive feature of the European way to philosophy of science. Now, the word 'pluralism' appeared in the programmatic leaflet of the Network, and the very choice of the speakers was guided by a pluralistic attitude. The Co-ordinating Committee wanted a wide range of viewpoints to be represented at the workshops, but a far better result was obtained, as a whole cluster of pluralistic approaches and solutions to the problems at stake were outlined there. To mention but a couple of examples, the reader is addressed to the papers given at the Vienna workshop by Theo Kuipers and Ilkka Niiniluoto, who showed that the traditional opposition between inductivism and deductivism can (and should) give way to a more eclectic standpoint.

A further characteristic of the European approach to philosophy of science lies with its historical concern. As the fathers of philosophy of science—from John Herschel to Pierre Duhem, Moritz Schlick, Otto Neurath and many others—taught us, the awareness of the historical depth of problems and concepts is always a good guide to detect their interconnections with other problems and concepts. Such an awareness is a hallmark of European philosophy of science, and inspires a typically 'global' approach to problems, according to which even when a particular problem is addressed, this is done with an eye on the wider debate (past and present).

After a period of 'exile' in the New World, philosophy of science is again flourishing in Europe, and hopefully the work carried out by the Network has contributed an epsilon to it. To be sure, in a time of globalization this claim should not be taken to suggest barren and anachronistic oppositions. But it is desirable that the European tradition and approach to philosophy of science will be strengthened further and further, as this can only result in an enrichment of this discipline.

To conclude this report, it seems appropriate to mention the other members of the Coordinating Committee of the Network on 'Historical and contemporary perspectives of philosophy of science in Europe': Aristides Baltas (National Technical University, Athens), Donald Gillies (King's College, London), Theo Kuipers (University of Groningen), Ilkka Niiniluoto (University of Helsinki), Michel Paty (University of Paris 7), Miklos Redei (Lorand Eotvos University in Budapest), Friedrich Stadler (University of Vienna and Institute Vienna Circle), Gereon Wolters (University of Konstanz). As the Chairperson of the Coordinating Committee, I wish to thank them all for their wonderful cooperation. On behalf of the whole Committee, I also wish to thank all the speakers who accepted to take part in the workshops organized by the Network, and have made them valuable occasions for doing philosophy of science, the European way. Last but not least, warmest thanks are due to the European Science Foundation for supporting our work.

Maria Carla Galavotti

Department of Philosophy
University of Bologna
via Zamboni 38
40126 Bologna
Italy
Email: galavott@philo.unibo.it

INDEX

elson, R., 19
clic, 190
ámek, J., 149, 153
blard, F., 20
dler, D., 248
glo-American philosophy, 139
malous monism, viii, 115
scombe, E., 81
lication innovation, 15
lication-dominated research, vi,
    4, 6–8, 10
umentation framework, 197
umentation network, 197, 224,
    239–241
  recursive, 198
ignment, 177
kinson, D., 248, 249
om of choice, 149
om of determination, 149

kpropagation learning algorithm,
    240
ns, W., 8
tas, A., 248, 251
nzhaf (or Coleman–Banzhaf) in-
    dex, 48
nzhaf, J., 48
ringer, H., 219
tterman, R., 12, 13
ich, B., 140
yesian, 94–97
yesian multinet, 193
  recursive, 193, 194
yesian network, vi, xi, xii, 173,
      176–180, 182, 184, 186,
      187, 189, 190, 192–197,
      199–201, 203, 209, 219,

      223–225, 228–230, 233–
      235, 237–239, 241
  hierarchical, 196
  object-oriented, 195
  recursive, 173, 179, 180, 187,
      189–193, 195–199, 219, 223,
      224, 227, 234, 238
  recursive relational, 194
  relational, 194
BC-model, x, 20, 37, 41, 42, 45
Belief-Desire-Intention model, 228,
    229
Benacerraf, P., 168, 169
Bench-Capon, T., 197–199, 241
Bennett, J., 81, 225
betting quotient, 94
Bevan, K., 12
Block, N., 120
Bonjour, L., 90, 91
Boolean algebra, 149, 150, 152
bounded confidence model, *see* BC-
    model, 42
Briancon, C., 145
bridge law, 117
bridge laws, ix, 118
Brier rules, 96
British idealism, 139
Broome, J., 83
Buchanan, J., 111
Butterfield, J., 128, 249

C ampaner, R., vii
Campaner, R., 248
Carnap, R., 140, 146
Carrier, M., vi, 6, 8, 15
Cartwright, N., 1–3, 9, 127, 156
Cassirer's Critical Idealism, ix

Cassirer, E., ix–xi, 139–146, 148–151, 153–158, 161–170
causal chain, 173, 174, 176, 181, 185, 187, 229
causal dependence condition, 184
causal graph, 174, 177, 178, 181, 182, 193
   recursive, 174
causal irrelevance principle, 193
causal model, xi, 173, 196, 197
causal relation, xi, 173, 174, 177, 178, 185, 199, 223, 225–228, 231, 234
causal republicanism, vii, 76
causal supergraph, 181
causal supernet, 184
causal-functional account of truth-directedness, 80, 104
causality, 174–176, 194, 197, 219, 223–225, 227, 230
   recursive, xi, 173, 174, 176, 178, 194, 196, 197, 223, 225–228, 230, 231, 234
causally consistent, 181, 182
Causey, R., 131, 133
chain, 181
Chatterjee, S., 20
Clark, P., 248, 249
closest common cause, 182–185, 187, 188, 224
coefficient of variation, 23, 24, 41
Coffa, A., 140
Cohen, H., 169
Coleman, J., 48, 197
collective decision making, 47
common cause, 184
common foundational syntheses, 143, 144
common foundational synthesis, ix
concretization, 121–124, 128
consensus, 20, 21, 24, 29, 31, 38
consistency, 180, 182, 187, 188, 197, 223, 224, 234

causal, 180–183, 187, 196, 197, 223, 234
   Markov, 180, 182–184, 186, 187, 196, 223, 224, 234
   probabilistic, 181, 184, 187, 196, 223, 234
consistent, 187
contextualized causal relation, 6, 8
continental drift, 97
coordinating definition, viii, 117, 118, 121, 132
Copernican revolution, 74
Copp, D., 81
Corfield, D., 145
corrective function, 123
critical idealism, x, 139, 140, 142, 145, 157, 161, 162, 170
cut rule, 213, 216

D-separates, 183
Davey, B., 150, 153
David, M., 93
Davidson, D., viii, 115, 118, 119
De Groot, M., 19
Dedekind cut, ix, 146, 153, 163
Dedekind, R., x, 148, 150, 152, 153
Deffuant, G., 20
degree of belief, 94, 96, 97
Dempster–Shafer, 202, 209, 235
deontic logic, 233
depth, 179
Descartes, R., 73, 76, 88
design rules, 7
direct inferior, 179
direct superior, 179
directed acyclic graph, 177, 181
directional function, 123
Dollar/Euro parity, viii, 108, 109
Douven, I., 248
Doxastic, 85
dualism, 120, 127
Dung, P. M., 197

utch book, 94, 96

ngel, P., vi–viii, 42, 83, 85, 88,
        101–104, 107–111
scher, M., 48
sential incompatibility of theory,
        128
sential structure, 122, 123
,sentially compatible theory, 124
,sentially incompatible theory, viii,
        ix, 124, 126, 137
uler's theorem, 156
,ent-theorist, 227
,actness problem, 168, 169
,plication, 117
,tensive abstraction, 152

,ctual statement, 123
,edforward network, 236, 237
,lsenthal, D., 48
,sta, R., 248
,red network, 234, 244
,ite, 179
,ach, P., 196
,attening, 178, 189, 197
,dor, J., 119, 120
,rmal decision theory, 115
,rmalism, 145
,ench, J., 19
,iedman, M., 140, 161, 164, 165
,ndamental theorem of algebra,
        147
,sion reactor, 10, 11

,aa, J., 117
,abbay, D., vi, xi, xii, 249
,abbay, D. M., 219, 223, 226–230,
        232–234, 238
,alavotti, M., v, ix, xiii, 248
,arbolino, P., 248
,arcez, A., xii, 219
,nerator, 32
,ettier's problem, 92
,ibbard, A., 96, 97

Giere, R., 9, 166, 167
Gigerenzer, G., 248
Gillies, D., v, 47, 248, 251
Glass, D., xi, 219
Gonzalez, W., 248
Gozzano, S., 98
Grattan-Guinness, I., 248, 249
Grobler, A., 248
Gyftodimos, E., 196

Haack, S., 139, 140, 156
Hacking, I., 9
Hacohen, M., 248
Halpern, J. Y., 231
Hamilton's quaternions, 156
Harary, F., 19
Hartmann, S., 1, 12, 248
Hayek, F., 111
Heal, J., 91
Hegselmann, R., vi, x, xi, 19, 20,
        24, 26, 29, 31, 47, 249
Hempel, C., 146
Herrlich, H., 148
Hon, G., 248
Honderich, T., 116, 118
Howson, C., 248
Hoyingen-Huene, P., 170
Hull, D., 134
Humberstone, L., 81, 85

ideal element, ix, x, 143–147, 149,
        150, 156, 163–165
ideal gas law, viii, 155
idealizataional law, 123
idealizational conception of science,
        viii, ix, 115, 116, 121, 122,
        127, 135
idealizational law, viii, 121–124,
        126
idealizing assumption, 123
incommensurability, 117, 170
inferior, 179
information network, 199, 200
        self-fibred, 219

self-fibring, 173
input, 199
input-output, 199
instantiated, 177
intentional account of truth-directedness, 86
interior, 181
intuitionistic logic, 209
Irzik, G., 248
iterated weighted averaging, 20

Jackson, F., 84
Jaeger, M., 194, 195
Jaynes, G. T., 192
Jeffrey, R., 95
Johnston, M., 73, 74
Joyce, J., 95–97

Köhler, E., 248
Kant, I., vii, 74, 75, 141, 161, 162, 164, 168
Keller, E., 8
Kepler, J., 154
Keynes, M., xi
Keynesian revolution, 113
Kim, J., 116, 118, 120, 127, 134, 226, 227, 232
KISS-principle, 20
Kitcher, P., 5, 167, 168
Klein, U., 248
Koller, D., 195
Krajewski, W., 121
Krause, U., 19, 20, 24, 26, 29, 31
Kripke, S., 132
Krois, J., 161
Kuhn, T., 117, 139, 170, 250
Kuipers, T., 122, 124, 128, 248, 249, 251
Kvasz, L., 248

Łastowski, K., 128
labelled deductive system, 202, 213, 238
labour market, 112

Lakatos, I., x, 156
Lamb, L., 219
Lambek Calculus, 214, 215
Lambek substitution, 216
Lauritzen, S. L., 200
laws of celestial mechanics, 1
Legrenzi, P., 248
Lehrer, K., 20, 41, 43
level 0, 179
level 1, 179
level 2, 179
Levi, I., 98
Lewis, D., 117, 230
linear logic, 209
linear space logic, 233
Linsmeier, K.-D., 6
liquid drop model, 3, 11
local model, vi, 5, 6, 8, 10–14
locality, 33
logical empiricism, 140
logicism, 145
Longino, H., 8
Lorenz, J., 37
Lucas, R., 111, 112

Machover, M., xi, 47
magnetic read head, vi, 7
Marburg school, 165
market economy, 110, 111
Markov
    consistency, 182
Markov condition, 173, 178
    causal, 177, 178, 180, 182–184, 189–193, 197, 224
    recursive, 189–191, 193, 196, 224
Marx, K., 112
materialism, 115, 116, 120, 127
mathematical induction, x, 169
mathematisation, 139–141
maximum entropy principle, 193
McDowell, J., 119
McLaughlin, B., 116, 118
Meheus, J., 248, 249

ellor, D., 102, 174
enzies, P., 74
erin, A., 98
eta-model, 73
etaphysics, 223, 225
ethodological deficiency, 4, 7, 10, 11, 14, 15
ilford, K., 248, 249
iller, D., 248
ind-to-world direction of fit, 81
inimal model, 12, 13, 19, 44
odel of the solar system, 1
odels of a theory, 1, 2
oderate conventionalism, x, 157, 161, 167
onetarism, 113
oore's paradox, 81
oore, G., 139
organ, M., 2
ormann, T., ix, x, 152, 161–165, 167
orrison, M., 2, 3, 9, 10

agel, E., ix, 117, 121, 133, 134
anoresearch, 7
avier-Stokes equation, vi, 10, 12
eal, R. M., 190
eapolitan, R. E., 178, 192
eau, D., 20
eo-Kantian idealism, 144
eo-Kantian idealist, 140
etwork argument, 198
etwork variable, 178
eural network, xii, 207, 233–238, 240, 242, 243
    argumentation, 240
    fibred, 236
    recursive, 236–238
eural networks, vi
eural-symbolic learning system, 238, 242
iiniluoto, I., 248, 249, 251
on-recursive, 179
oordhof, P., 80, 87

Nordmann, A., 7, 15
norm of truth for belief, 83, 84, 94
normative account of truth direct-
    edness, 83
norms of rationality, 89
Nowak, L., ix, 116, 121, 122, 126, 128, 135, 143
Nowakowa, I., 121, 143

object, 195
objective probability, 73
Olsson, E., 98
opinion dependent asymmetry, 29, 30
opinion dynamics, x, 19, 21, 33, 37, 41, 43, 45
opinion independent asymmetry, 27
opinion space, 21, 23, 29–31, 33, 35, 37, 38, 41
opinion world, 25
optical switch, 6
output, 199
Owens, D., 87, 89, 90

pair-wise sequential updating, 35
pair-wise updating, 35
Papineau, D., 248
Paprzycka, K., vi, viii, ix, 117, 121, 122, 131–136
Paprzycki, M., 122, 136
Pascal, B., 145
Passmore, J., 161
path, 181
Paty, M., 251
Pauli's exclusion principle, 3
Peacocke, C., 92
Peano's axiomatisation, 169
Peano, G., 169
Pearl, J., 176, 178, 183, 191, 192, 199, 231
peers, 179
Peijnenburg, J., 248
Peirce, C., 88

258

Peña, J., 193, 194
Penrose, L., xi, 48
Penrose, R., 48
Pfeffer, A., 195
phenomenological law, 2, 127
phenomenological model, 2
philosophy of symbolic form, 161
physicalism, 119
Pickering, A., 156
plurality, 24, 26, 29, 33
Poincaré, H., x, 161, 167–170
polarisation, 26, 29, 31, 33, 41
polarity, 24, 25
practical impossibility problem, 167
Prandtl, L., vi, 10–12
Price, H., vi, vii, 73–77, 98
Priestley, H., 150, 153
principle of the nomological character of causality, 119
probabilistically consistent, 184
probability specification, 177
projectivist metaphor, 73
proteomics, 8, 15
Psillos, S., 248
Putnam, H., 119, 168, 169

Rabinowicz, W., vii, 98, 128
radically simplifying model, x, 19, 43
Ramsey, F., 94, 95, 102
Ramsey, J., 2
random mean, 38–41
real mathematics, 143, 145
recursive Bayesian network, xi
recursive causality, 223
recursive structural equation model, 197
Redei, M., 248, 251
reductionism, vi, 115–117, 121, 127, 128, 131, 133
Reed, M., 7
Reichenbach, H., 250
republican perspectivalism, 73
Richardson, A., 140, 144

Rickert, H., 140
Rorty, R., 91
Rosner, P., vii, viii, 98
Rott, H., 248, 249
Russell, B., vi, vii, 73, 76, 139, 161

Sahlin, N., 248
Salmon, W., 5, 135
sameness thesis, 142, 161–163
Sartwell, C., 93
Scazzieri, R., 248
Schaffer, J., 225, 231
Schaffner, K., 121
Scheffler, I., 139
Schoenle, P., 248
Schweber, S., 4
Searle, J., 81
self-fibred, 200
self-fibred network, 203
self-fibring network, xi, 199, 211, 223, 224, 234
Selten, R., 248
SEM-variables, 197
semantic insertion, 201, 203, 235, 238
semigroup, 205
Shallice, T., 249
shell model, 3, 11
Shrader-Frechett, K., 12
simple argument, 198
simple variable, 178
simplification, 180
simultaneous updating, 26, 35
Sintonen, M., 248
Smith, D., 145
Smith, M., 81
Sobel, D., 81
social choice, 47
social constructivism, 42
Spiegelhalter, D. J., 200
Spirtes, P., 190
Spohn, W., 98, 248, 249
Stöckler, M., 1

ltzner, M., 10, 11, 248
dler, F., 248, 249, 251
lnaker, R., 81, 84
udt, K. von, 146
glitz, J., 111, 112
ne's representation theorem, 148, 149
ne, M., 150
utland, F., viii, 116, 118
ct implication, 214, 215
ct substitution, 216
ictural equation model, xi, 173, 196, 197, 219
  recursive, 197, 219
ictural realist, 167
chain, 181
jective probability, 110
erior, 179
pes, P., 248–250
tactical substitution, 201, 203
bo, L., 248, 249
zeciniarz, J., 248

penden, J., 146
iporal logic, 233
en-token identity, viii, 115, 118, 128
retti, R., 145, 146
ir, J., 7
nscendental idealism, 140
th directedness of belief, vii
th-directedness of belief, 79, 80, 84, 101
e-epiphenomenalism, viii, 115, 116, 118, 119, 127
e-type identity, viii, ix, 115, 118– 120, 127, 131–134

hs, M., 248

Dyck, M., x
leman, D., vii, 83, 85–88, 102– 105
leman, O., 79, 80

Voronoi diagram, 32, 33
voting power, 48
voting procedures, 47

Waals, J. van der, viii, 122, 124
Wagner, C., 20, 41, 43
Walker, T., 79, 87
Waters, C., 134
Weber, E., x
Weber, M., ix, 128, 131, 135
Wedgewood, R., 83, 92
Weisbuch, G., 20
well-founded, 179
Whitehead, A., ix, 152
Wilholt, T., 7, 15
Williams, B., 82, 84–86, 110
Williamson, J., vi, xi, 178, 192, 193, 223, 226–230, 232, 234
Williamson, T., 92, 93
Wilson, M., 144–146, 149, 156
Winsberg, E., 2, 9, 11
Winters, B., 85
Wittgenstein, L., 81
Wojcicki, R., 248
Wolters, G., 98, 248, 251
Woods, J., 219, 238
world-to-mind direction of fit, 81, 82, 88
Wright, C., 140

Zangwill, N., 83